Responsible AI

Sray Agarwal • Shashin Mishra

Responsible AI

Implementing Ethical and Unbiased Algorithms

Sray Agarwal
London, UK

Shashin Mishra
London, UK

ISBN 978-3-030-76859-1   ISBN 978-3-030-76860-7   (eBook)
https://doi.org/10.1007/978-3-030-76860-7

This Springer imprint is published by the registered company Springer Nature Switzerland AG
The registered company address is: Gewerbestrasse 11, 6330 Cham, Switzerland

Foreword

"Artificial Intelligence is everywhere". This sentence has become relatively common in our technological societies. Yet it is misleading for many reasons. First, because the access to such technology is uneven at both the international and domestic levels. Some areas of the world are either not equipped with or do not have a correct access to artificial intelligence systems and infrastructures (AIS&I), and in many countries, developed or not, some people do not have access to AIS&I. So technically artificial intelligence (AI) is not everywhere.

Second, to state firmly that AI is everywhere, request first and foremost a clear definition of AI. Short of it, AI is just a phrase, a notion, an idea translated into a narrative. This point is certainly the most pressing issue we are facing with AI. Like a new deity, AI exists only through the assertion of its existence. It is there without being; therefore, paradoxically, we talk about it, have an opinion on it, but are unable to define it, or at least agree on a unique definition.

Thus, the fuzziness of the notion has opened the doors to all kind of phantasms. For some, it appears as an existential threat that needs to be strictly controlled or even stopped. For others, it carries the promise of significant progress that would benefit humanity. Anyhow, and whatever the stance, AI is referred to as if it has its own existence outside of any strict explanation.

There are many available definitions of AI. None of them is perfect. None of them has reached a consensus. From Alan Turing or John MacCarthy's definitions of AI as "thinking machines" to Marvin Minsky writing that AI is "the science of making machines do things that would require intelligence if done by men", encompassing "high-level mental processes such as: perceptual learning, memory and critical thinking", to more recent definitions provided by private or public organizations, it seems that artificial intelligence basically refers to systems that, thanks to algorithms, are able to mimic some human brain abilities such as learning, reasoning, planning, problem-solving or identifying patterns.

The problem here is that all these definitions are themselves made of words that can be attributed different meanings. As the French linguist Ferdinand de Saussure demonstrated it, there is a huge difference between the signified, the word as we hear or read it, and the signifier, the meaning we assign it, that together constitute a sign, namely a word. Language cannot then be understood outside of its social environment. Consequently, each and every word we use when we talk about AI can be subject to several definitions which in turn would shape in different ways our perceptions on the subject. As Ludwig Wittgenstein asserted it in his *Tractatus Logico-Philosophicus,* "*[t]he limits of my language* mean the limits of my world".

Thus, even before trying to take a stance on the risks and benefits of AI, or on the way to control it, we should ask ourselves how can we define artificial? What does intelligence refer to precisely? And eventually, what these two words stuck together mean? Obviously, definitions of intelligence and artificial are to be found in thesaurus. But, when it comes to technologies that could have a huge impact on humanity, thesaurus is not enough. And then philosophical questions arise. Looking at "artificial" in the Merriam-Webster thesaurus we can read that artificial means "humanly contrived often on a natural model". But what does "humanly" mean exactly? How can we define "human"? Is there any consensus on the way "human" is defined?

What about intelligence? Who can pretend knowing what intelligence is? Are we talking about emotional or situational intelligence? Is intelligence the mere collection and processing of data? Is it understanding, learning, reasoning, planning, problem-solving …? In AI does intelligence refer to several or all of its naturalist, spatial, musical, bodily-kinesthetic, logical-mathematical, interpersonal, intrapersonal, or linguistic dimensions identified by Howard Gardner? One could even ask whether intelligence, natural or not, exists.

Without any clear and common signifier all the words we are using in the field of AI can be understood in very extensive ways, and consequently to various interpretations, perceptions and behaviours. For constructivists, language is the key vehicle for ideas and meanings. It contributes to the shaping of our perceptions, our thoughts, our definition of things, and eventually the way we act. At the end of the day, depending on how one understands intelligence, AI can either represent a threat or a benefit. In his excellent book *AI Ethics*, Belgian philosopher of technology Mark Coeckelbergh relevantly reminds us that "our evaluation of AI seems to depend on what we think AI is and can become".

Then words must be used with the utmost precaution to avoid sending wrong messages or opening the door to multiple interpretations. Words can be weapons, and language can be a tyranny. Our tendency to use words and narrative without thorough preliminary reflection can lead to ideological perspectives and the polarization of stances instead of fostering a constructive debate. In their book, Shashin Mishra and Sray Agarwal are well aware of that risk. When they address the question of what they call "Privileged/unprivileged classes", they relevantly clarify that "a privileged class for one example may be unprivileged class for another" and that consequently, "[b]y transferring our understanding of privileged class from one problem to another we may end up compounding the bias and not actually reduce it". I would even add that we may even make the bias worse.

Hence, we must avoid polarization, subjective notions and loosely defined words. Nuances are key, human beings are made of nuances; they are not binary. This is exactly what we are (re)discovering thanks to AI. Trying to duplicate human intelligence, we are diving deep into the intricacies of human beings. If there is something we have learned from AI, it is that we do not know ourselves. We are ignorant about the very nature of humanness. Nothing new: philosophers have been debating on the subject for centuries.

Words can shape the world as we see it. Ill-defined words leave the door wide opened to all kind of fantasies. Thus, the mere idea that AI-fitted machine could reproduce brain's abilities suffices to raise concerns about the potential autonomy AI could reach. Indeed, if machines could at some point think by themselves and make decisions, how would they act towards human beings? Would they collaborate? Would they accept to remain as tools in our hands? Or, as Elon Musk and Stephen Hawking warned us, would they become an existential threat to humanity?

So far there is no clear answer to these questions. Only possibilities with different degrees of probability. One way to tentatively address this, would be to ask ourselves whether we will be able to keep control over AI.

And there is our second issue with AI: our human arrogance leads us to firmly believe that we are and will remain in control of AI, in other words of something we do not know about.

This is a long-lasting debate related to what Aristotle called *techné* or technique, understood as both an art and a process. A knowledge system that completes and imitates nature, according to the Stagirite. The desire to control our environment, to make our lives easier has led us to think that technique means progress, and that progress is good if not desirable. Technique has slowly turned into a tool aiming at increasing pleasure, living a "better life", albeit a good life, whatever these words mean.

So, we think that AI is nothing else but another technique, and that therefore it can be controlled as easily as a hammer. Humans will remain in the loop, or maybe on the loop, or out of the loop! But they will remain somewhere around to monitor and control AI for the benefit of humanity. What if we were wrong? What if we already have lost control?

Are we in control of techniques? Tough question. We tend to think we are. We would like to think we are. Yet, it seems that technology has, to some extent, already taken the lead over human beings. Our addictions to social network, our dependency to our smartphones, our reliance on the Internet, are inviting us to question the firm belief we have in our control over techniques. Techniques contribute to the shaping of the world we live in. Just think about cars, the way they changed our perceptions of distances, the way they influenced city settings, their impact on the way our environment is designed, the way they give rhythm to our everyday lives.

The myth of control is perfectly explained by the Canadian philosopher of technology, Andrew Feenberg. According to him, the instrumental approach postulates that techniques (or technologies) are neutral tools, that they basically serve users' purposes, and that they have no influence over societies. But Feenberg also explains that on the other hand, there is a substantivist approach that contrariwise affirms that techniques are shaping our cultural systems, that they are in fact autonomous enough to have their dynamics and to restructure our social environment. In other words, and as Gilbert Simondon put it in his seminal *On the Mode of Existence of Technical Objects*, technical objects have their own lives.

The myth of control is to be questioned. Our will to master nature, to free humanity from its flaws, or more precisely what is seen as flaws, must be questioned. Our belief that techniques are progress, and that progress is good and therefore desirable, must be questioned.

Sixty years ago, in *The Question Concerning Technology*, German philosopher Martin Heidegger already wrote that "[e]verywhere we remain unfree and chained to technology, whether we passionately affirm or deny it" and warned us that "the will to mastery becomes all the more urgent the more technology threatens to slip from human control".

The point is definitely not to fall into the trap of techno-scepticism and apocalyptic considerations. It is to warn against the belief that AI is something we know enough to keep tight control over it. The point is to invite to ask questions, relevant questions, instead of trying to offer answers and solutions.

In our hedonistic societies, individual well-being and satisfaction has become an end in itself, and we believe that AI will participate in the improvement of our lives freeing us from pain, from discrimination, from any kind of unfair treatment, from needs, from sufferings, nay from death.

We do, I think mistakenly, think that pain should be eradicated from the surface of the globe. We are convinced that the future of humanity should be made of AI systems that would be "more human than humans". Starting from there AI should be cleared out from human flaws. But let's ask some philosophical questions here: are humans without flaws still humans? Is perfection an end in itself? Does it exist? Isn't perfection precisely what human being already is, namely a subtle balance of qualities and flaws?

This book deals with AI and the way we could develop a responsible AI given that, and following the substantivist approach, it impacts our everyday lives sometimes in undesirable manners through "unfavourable outcomes" as Sray Agarwal and Shashin Mishra put it.

Doing so it stresses upon interesting issues like fairness and the way we should detect biases and discrimination in order to avoid them. It also deals with explainable AI, namely the need to develop AIS&I for which we can provide, when requested, explanation on data used, collection and processing, on models' performances and outputs. The solutions offered in this book are aiming at developing AI-fitted tools that would be beneficial for the greatest number, that would be freed from biases and potential drifts.

Is it possible? Is it desirable? There is no clear answer to these questions so far. Focused on solutions to supposedly existing potential problems related to a supposedly existing technique, namely AI, we are diverted from asking relevant questions such as "what kind of society are we aiming at?"

We suppose that human beings can both be mimicked and improved, if improving humans ever makes sense, by techniques. This perspective is no recent. In 1651, Thomas Hobbes stated at the very beginning of the introduction of his *Leviathan* that art "can make an artificial animal", that life is nothing else than "a motion of limbs", and that organs such as the heart, the nerves, the joints are mere spring, strings, and wheels. For the political philosopher, "[a]rt goes yet further, imitating that rational and most excellent work of Nature, man", creating artificial life. One century later, Julien Offray de la Mettrie, asserted that "man is a machine". Nowadays, we are discussing the advent of enhanced humans, kind of Nietzschean superhumans, we are thinking of trans-humans and post-humans made of flesh and steel. From the very idea that human beings are mere machines that can be reproduced by art (*technē*) to the idea that humans will mix with machines to the idea that machines are the future of humans, wide gaps have been passed, inviting us to wonder about what it means to be human and toward which kind of world we want to head to.

Is AI as a *technē*, aiming at completing and imitating nature as Aristotle wrote it? Is it aiming at creating new sentient beings either machines or humans-machines? Is it what we want? More than that, when we pretend to remove flaws such as biases from AI, do we want to transform the reality of our world or do we want to reproduce it? If we want to transform it, towards which model do we want to go and to which extent this model would work for the entire humankind? Is it even ethically acceptable or desirable to transform the reality? Then, reading the following pages, one should wonder what "reweighting the data" entails? What reducing "the bias or the discrimination in the data" means exactly? Applying principles is not enough. Applying them relevantly and with full knowledge of the facts, is key.

Obviously, discrimination can be detrimental in some cases. Obviously, whenever relevant biases that can harm should be mitigated and, if possible, removed. Does that mean that all discrimination is to be abolished? By whom? On which basis?

It is important to take this book and its content for what it is: a reflection on specific issues related to specific situations. Any attempt to universalize fairness, trustworthiness, the abolition of discrimination and so on, is risky for it would deny the diversity of perspectives and the contingent nature of values.

Does that not mean that offering potential solutions to make sure AI is designed, developed, and used in a fair way, is pointless? Quite the opposite. Every reflection is welcome as long as it is applied in specific cases without any universal ambition. In their book, Shashin Mishra and Sray Agarwal are providing us with qualitative and insightful thoughts regarding fairness and responsibility. They do not pretend offering cure-all solutions. It is the responsibility of the reader to evaluate the relevance of the options proposed by the authors regarding their own situation. No doubt lots of ideas and solutions shared in the coming pages will help in many cases if used accordingly.

Nonetheless, we might be cautious not to focus on AI at the expense of human beings. We must be careful not to reify AI, giving it attributes it cannot have. I have already mentioned the problem of words and their definitions. It is important to avoid applying virtues to AI instead of humans. "Trustworthy AI" and "responsible AI" are two examples of this tropism. AI cannot be trusted, not to mention that it should not. Trust is relation with others, not with objects. AI must not be the subject of trust, but human beings that are designing it, programming it, developing it and using it must be. As German philosopher Professor Thomas Metzinger, who was a member of the commission's expert group that has worked on the *European ethics guidelines for artificial intelligence*, asserted, "[t]he Trustworthy AI story is a marketing narrative". The responsibility that goes with the potential sanctions does not lie with the AI and is instead measured in regards of the agent's intentions. Unless AI becomes autonomous in a Kantian sense, that is it can express some kind of volition, it cannot be held responsible, but human beings that are designing it, programming it, developing it and using it must be.

Finally, I would stress that any work aiming at fostering the debate or bringing new ideas in the field of AI must be embraced. Each work, this book included, is a stone participating in the construction of a bigger edifice. It is up to us, individually and collectively, to make sure these stones are used and sufficiently embedded in each other to build something strong and durable, following Hans Jonas *The Imperative of Responsibility*: "Act so that the effects of your action are compatible with the permanence of genuine human life". I respect and praise the fact that Shashin Mishra and Sray Agarwal are taking their responsibilities participating in the reflection on AI.

A book like this one is not a panacea, it is a step, a huge one undoubtedly, but a mere step. It is up to us all to take this step and after the next to move towards something we want, the way we want. But before that, we need to ask essential, complex and numerous questions.

I am convinced that Sray Agarwal and Shashin Mishra have done a useful and thorough work, providing all people working in the field of AI valuable food for thought and opening ways to potential solutions to specific problems. In doing so, they have brought a huge stone on the AI work site.

As you will notice, this book is highly technical. I am deeply honoured and grateful Shashin Mishra and Sray Agarwal for inviting me to complement their incredible work with my humble and limited perspective from human sciences.

I am sure you will enjoy this book as much as I enjoyed it. I wish you good reading and fruitful reflections.

Dr Emmanuel R. Goffi, is a philosopher of techniques and an AI ethicist. He currently is the co-director and co-founder of The Global AI Ethics Institute. He is also a research associate with the Big Data Lab at the Goethe Universität in Frankfurt Germany, and with the Centre for Defence and Security Studies at the University of Manitoba in Winnipeg, Canada.

Paris, 2021   Emmanuel R. Goffi

Preface

The need for AI to be responsible has never been more urgent than it is now. There's hardly any aspect of our lives now that is not touched by the technology. However, most of the AI-driven products fall short on more than one aspect of responsible AI, and the reason most often is the lack of structured information on why to do it and how.

Most of the books that go deep on the subject matter seem to target data scientists only. To make a robust software product, other roles in the team, including the business sponsors, are equally important, and an AI implementation is no different. This encouraged us to write a book that aims to help all the roles within a product team.

We hope you enjoy reading and applying the concepts from this book as much as we enjoyed writing it.

Who is this Book for?

This book is for all! Anyone who is working towards building an AI-driven product can benefit from this book. If you are sponsoring products for your line of business, this book can help you understand how your AI products need to follow ethical AI practices and what do these mean for your consumers as well as the bottom line.

A product owner needs to ask the right questions and ensure that their team is able to implement the responsible AI while building the product that the business requires. This book can help you understand the fundamentals behind RAI and help you make the right decisions.

A business analyst needs to be able to understand the inner workings more closely than the product owner and has to work with data engineers and data scientists in the team to choose the algorithms that need to be implemented as well as the metrics for measuring the impact. This book will take them through some math and the basics behind the different techniques so that they can not only use them to make decisions but also help the business understand them.

Finally, the book is also designed to be a handbook for data scientists at all career stages. The book proposes several new techniques and methods for implementing various components of RAI, along with the sample source code and a code repository with more supplementary code, including a notebook for each chapter.

How to Read this Book

Even though the book follows the lifecycle of a model, except for the Data & Model Privacy, you don't need to follow the book in this sequence, unless you are looking at solving a problem from ground up. We have a running use case across the book (except for a few parts) on the open-sourced data set extracted from public loan book from Bondora, a P2P lending platform. However, the hands-on codes and concepts can be easily implemented for any use case across industry.

Having said that, there is no particular sequence that needs to be followed. While a data scientist would be interested in reading the book cover to cover, a PO and BA would find chapters on proxy features, fairness definitions and model accountability more relevant and relatable to start with. The book also contains step-by-step python codes that will help you understand the implementation and intuitions much better. However, the supplementary code repository has more complex and optimized code meant for production and reusability.

How to Access the Code

This book is accompanied with source code for most of the techniques covered. Each chapter with code has an accompanying notebook. As far as possible, we have used the same dataset throughout the book to show how different techniques can be applied for a given problem and dataset. The common dataset is available in its own location within the code repository. If there is a chapter-specific dataset, it is available within the chapter's folder in the repository. The QR code below contains instructions on how to access the source code, following a short questionnaire.

London, UK

Sray Agarwal
Shashin Mishra

Acknowledgements

We would like to thank our family members for allowing us to use family time, for over a year, towards the book. This book wouldn't have been possible the without unparalleled love and support of our spouses (and children), and the encouragement from our parents and siblings.

We would also like to thank our colleagues in Publicis Sapient and clients whom we have worked with over the years. The work done on numerous projects together has been a learning experience, and their support, deliberation, and questions helped us give a comprehensive and concrete shape to the book.

We would like to especially thank Dr Emmanuel Goffi, co-director and co-founder at Global AI Ethics Institute in Paris, for taking out time from his busy schedule to write a foreword to this book, which gives the reader a lot to think about.

We would also like to thank our editor Paul Drougas and project coordinator Arun Siva Shanmugam for their patience and support throughout.

And finally, we would like to thank all the tireless researchers who are pushing ahead the boundaries of the responsible AI domain by introducing new methods and algorithms that meet the demands of the real world in performance as well as ethical parameters.

Shashin Mishra	Sray Agarwal
London, 2021	London, 2021

Contents

Chapter 1
Introduction

The machine ethics were mostly a topic for science fiction before the twenty-first century. The easy access and low cost of data storage and computing capabilities have meant that the use of machine-driven intelligence has increased significantly in the last two decades and the topic is not one for entertainment or a theoretical one anymore. The machine learning algorithms or AI now impacts our lives in innumerable ways – making decisions that can have material impact on the lives of the individuals affected by these decisions.

Through the year 2020, due to the global pandemic, even the sectors that shied away from technology have embraced it. However, the adoption has not been without problems, and as the number of lives touched by it have increased, so has the impact – positive or not. An example from very close to our homes is the grading fiasco for the A-level students (equivalent to the high school). An algorithm was used in the UK to predict grades that the students were likely to achieve as the examinations were cancelled due to the pandemic. The predicted grades were nowhere close to the centre-assessed grades (CAG), the grades that the schools, college or exam centre predict for the students. The CAG is used by the universities in the UK to give provisional places to the students as they are a reliable indicator of what the student is likely to achieve. With predicted grades far off from the CAG, many students lost their places in the universities.

It was later found that the algorithm, in order to fit the curve, increased grades for the students at small private schools, which are mostly around affluent neighbourhoods, and lowered it for large public schools, where a lot of students belong to ethnic minorities or from families with poor socio-economic status. This was quickly acknowledged as algorithmic bias, the algorithm was decommissioned, and CAG were awarded as final grades, but for a lot of students the damage was already done.

The problem of bias and lack of fairness is not limited to the first time, built in a rush model alone. A large number of discriminatory algorithms can simply go undetected as the impacted user group may be concentrated in a small geographical area as compared to the overall area targeted by the model, take, for instance, the price of The Princeton Review's online SAT tutoring packages. The price ranged from as low as $6600 to as high as $8400 based on the zip code of the user. Interestingly, the zip codes with the higher prices were also the ones with Asian majority population, including even the lower income neighbourhoods.

There are numerous examples of algorithms (and by extension products) that fail in implementing fairness, removing bias from their decisions or properly explaining these decisions. The reason these algorithms, or any algorithm for that matter, fail is because the facets of responsible AI are not considered when the data is being analysed, the objective function being defined, or the model pipeline being set up.

A big reason such products exist is because there aren't any resources that help identify the different facets to creating a responsible product and the pitfalls to avoid. Another challenge that the teams face is that the role of product managers/ owners and business analysts in the lifecycle of the model is not well defined. Our experience has shown us that all roles within a product team contribute to building an effective AI product, and, in this book, we have tried to cover the content from the point of view of all the roles and not just the data engineers and the data scientists. The book does get in detail, but wherever we have introduced math, we have tried to explain the concept as well as why is it necessary to understand. Let's begin by defining what is responsible AI.

What Is Responsible AI

The goals of responsible AI are manifold. It should be able to take decisions that reward users based on their achievements (approve credit for someone with good income and credit history) and should not discriminate against someone because of data attributes that are not in user control (reject credit for someone with good income and credit who lives in an otherwise poorer neighbourhood). It should be usable in a system built for the future with high levels of fairness and freedom from bias and yet should allow positive discrimination to correct the wrongs of the past. These requirements can be contradictory, and to overcome the inherent contradiction and build an AI that does this is what we want to talk about.

As the awareness is rising around creating responsible AI products, the actual implementations and the regulatory requirements, if any, are still catching up. In April 2020, the US Federal Trade Commission, in their blog titled "Using Artificial Intelligence and Algorithms", discussed why an algorithm needs to be transparent and discrimination-free. But transparency alone is not sufficient until it is accompanied with explainability, e.g. a 3-digit credit score is not helpful for the user unless it also tells them why they have a certain score and the actions that will help them improve the score. Just the transparency alone hides more than it reveals. The "Ethics guidelines for trustworthy AI" as published by the European Union in 2019 goes further and has defined a much wider framework.

Most of the time, the teams building AI-driven products are more concerned about what their products can do rather than how they do it. This results in biases creeping in and unfair decisions made by the systems, ultimately impacting the reputation of the product. AI done correctly would cover the different components of Responsible AI too – it is not just good for the user; it is good business too.

In this book we are presenting the different facets of responsible AI. The book aims to cover why these are important, what impact can they have on the model if ignored and the various techniques to achieve them.

Facets of Responsible AI

Depending on who you ask, the definition of what should constitute responsible AI (RAI) changes. This is primarily because even though there have been multiple works that call for the need for RAI, there has not been an end-to-end guide that covers the different facets and explains how to achieve them for your product. This book is our contribution to help make that possible.

As we take you through the data science lifecycle for one example, we have covered the actions that can be taken at each step to implement responsible AI, which in turn is divided into four parts – fairness, explainability, accountability and data and model privacy.

Figure 1.1 shows RAI components at each stage of the model's lifecycle. As you can see, the actions at each stage are different and complement each other to help implement an end-to-end RAI. Skipping over the components can leave the door open for the issues to creep in.

Let's have a closer look at each component of the responsible AI.

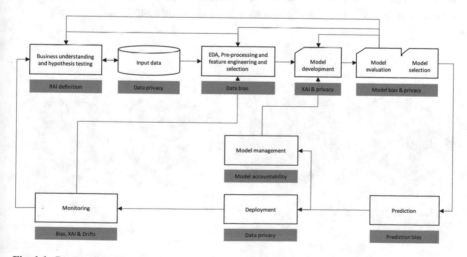

Fig. 1.1 Responsible AI components at different stages of the data science lifecycle

Fair AI

The biases in decision-making have been around since time immemorial and have had an influence on everything we do – for good or bad. The technology to make machine-driven decisions using algorithms is also not immune from our biases and can learn our biases, only they are incapable of seeing that as a bias and instead would consider them a gospel truth for the problem that the algorithm has been trained on.

In the early phase of ML adoption, removal of bias was usually not a concern, primarily due to lack of awareness of the issue. However, as different cases came to be known, the challenge in removing bias has mostly been the lack of understanding of how to do it. The hardest part in reducing bias and increasing fairness is the detection of the bias. Once detected, the urgency to remove it follows through automatically. The first place to check when your models show bias against a user group is to evaluate the data for historical bias, skewed representation of the user groups and for proxy features that got left behind undetected. In some instances, biases that were not even present in the original data can get introduced through engineered features, and in others, engineered features can hide or mask the discrimination that a model can still learn, making it even more difficult to detect the biases.

This makes it imperative to utilize statistical methods to check and mitigate biases in the data as part of the exploratory analysis. The pursuit of fairness continues after the model development, and the biases need to be checked regularly. Changes in the socio-economic fabric and underlying demographics can shift the fairness measures drastically. For example, the data may be unfair to a group initially but may gradually become unfair towards another group over time.

The chapters on fairness cover the possible use cases seen in the real-life scenarios and discusses solutions to detect and mitigate bias using the different components of fairness. Figure 1.2 shows the different components of fair AI that we should aim to cover as we work to making our algorithm more fair.

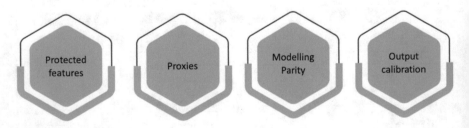

Fig. 1.2 Components of fairness

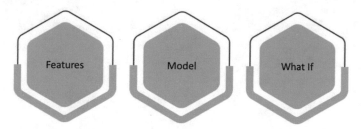

Fig. 1.3 Components of XAI

Explainable AI

Model explainability is a founding block for responsible AI and goes much further than the model transparency. Making a black box transparent by disclosing all the nuts and bolts won't do anything for responsible AI principles; however, adding ability to explain the decisions made by the model goes miles further than that in gaining user confidence and showing what is happening behind the scenes.

Explainable AI (XAI) needs to be designed keeping in mind the business, end users, stakeholders and regulating committee. Merely explaining the workings of the model isn't enough. You need to consider the context of the user as they will interact with your model and the type of explanation that will help them understand the model's behaviour. The three dimensions of XAI that we need to utilize to improve the explainability are shown in Fig. 1.3.

In the book, we will discuss not just the techniques but also the type of explanation needed. Given the fact that decisions and decision-makers need to see the problem from multiple dimensions, the coverage of explainability also needs to be multi-dimensional.

Accountable AI

Taking a model live is only the first half of the model's lifecycle. The second half involves staying live and relevant as the data distribution for the model changes over the period of time. Some form of model monitoring is often employed to track the data distribution; however, they are rarely used for monitoring the changes in the model's performance on responsible AI metrics. As a result, the model's performance on RAI can deteriorate significantly.

Monitoring should be accountable for detecting (and alarming) any deviation in explainability (feature, model and output) the way it creates alarms when there is data or accuracy deviation. It should further raise alarms for privacy deviations and show if privacy parameters need to be refreshed. It should also be used for monitoring fairness metrics and fairness measures. In the chapter on model accountability,

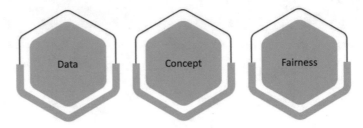

Fig. 1.4 Components of accountable AI

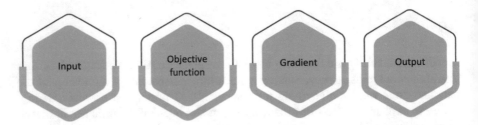

Fig. 1.5 Components of data and model privacy

we have discussed the primary duty of model monitoring (drifts and stability) but have extended the concept on how the same (or an altered version) should be used to monitor other founding blocks of responsible AI. Figure 1.4 shows the different dimensions of accountable AI that we will be covering later in the book.

Data and Model Privacy

Of the four facets of responsible AI, data and model privacy is perhaps the most important one and gets the most press too. User data privacy has had the regulatory support that other components of RAI have not. However, that is not always reflected in the real world. Studies have shown how attackers can reverse-engineer a model to find the training data, use data to reverse engineer the model or even poison the dataset influencing model's performance.

Thus, it's not only important to secure the data but also ensure the model being trained is secure too. A model that is prone to attacks cannot fair well in the responsible AI metrics. The components of data and model privacy are as shown in Fig. 1.5.

Within financial services industry, there have been quite a few discussions on creating regulatory sandbox, where companies can safely test their algorithm without being audited officially by regulators. The FCA and ICO have used regulatory sandboxes which was used by various firm for identity verification. This allows the firms to verify if their work is fair and inclusive. The efforts from the regulators and the industry continue, but there is still plenty of work to bring a standardized approach to responsible AI.

Within every product team utilizing AI to add intelligence to their products, the responsibility of "responsible AI" does not lie with a few stakeholders. All stakeholders have a role to play in there, from product owners and business analysts to the data scientists. Similarly, external stakeholders also influence the development of the standards and the adoption, like an ethics committee internal to the organization or a regulator.

Bibliography

Algorithms: How they can reduce competition and harm consumers (no date) Gov.uk. Available at: https://www.gov.uk/government/publications/algorithms-how-they-can-reduce-competition-and-harm-consumers/algorithms-how-they-can-reduce-competition-and-harm-consumers (Accessed: April 27, 2021).

Angwin, J., Mattu, S. and Larson, J. (no date) *The tiger mom tax: Asians are nearly twice as likely to get a higher price from Princeton review*, *Propublica.org*. Available at: https://www.propublica.org/article/asians-nearly-twice-as-likely-to-get-higher-price-from-princeton-review (Accessed: April 27, 2021).

"Blog: ICO regulatory sandbox" (2021). Available at: https://ico.org.uk/about-the-ico/news-and-events/news-and-blogs/2020/11/sandbox-helps-develop-innovative-tools-to-combat-financial-crime/?src=ilaw (Accessed: April 27, 2021).

Ethics guidelines for trustworthy AI (no date) *Europa.eu*. Available at: https://digital-strategy.ec.europa.eu/en/library/ethics-guidelines-trustworthy-ai (Accessed: April 27, 2021).

Harkness, T. (2020) *How Ofqual failed the algorithm test*, *Unherd.com*. Available at: https://unherd.com/2020/08/how-ofqual-failed-the-algorithm-test/ (Accessed: April 27, 2021).

Larson, J., Mattu, S. and Angwin, J. (no date) *Unintended consequences of geographic targeting*, *Propublica.org*. Available at: https://static.propublica.org/projects/princeton-review/princeton-review-methodology.pdf (Accessed: April 27, 2021).

Using artificial intelligence and algorithms (2020) *Ftc.gov*. Available at: https://www.ftc.gov/news-events/blogs/business-blog/2020/04/using-artificial-intelligence-algorithms (Accessed: April 27, 2021).

Chapter 2
Fairness and Proxy Features

Introduction

Humans have an innate sense of fairness, with studies showing that even 3-year-old children demonstrated ability to take merit into account when sharing rewards (ref: Young Children Consider Merit when Sharing Resources with Others). The natural ability to understand fairness is not limited just to the humans. Observational studies have shown that even primates respond to inequity, with reactions from chimpanzees mirroring those of humans when faced with inequity (ref: Nonhuman Species' Reactions to Inequity and their Implications for Fairness).

However, even though fairness is something we understand intuitively as a species, it doesn't necessarily always translate into our real-world experiences. Just as the fairness plays an important part in our decision-making process, so do our biases and prejudices – that can lead to discrimination. These biases and prejudices can be deliberate, like the discrimination based on skin colour of people, they can be based on stereotypes, like women not considered for jobs that require hard physical labour, or they can even be subconscious biases, that may reflect the practices or discriminations in place from the past generations. When observed within enterprise or business context, these biases can lead to discriminatory actions and can lead to one group of people receiving a favourable treatment (we will call them privileged class through the examples in this book) over other groups (we will call these unprivileged classes).

As businesses across industries incorporate machine learning as a part of their decision making, the historical actions captured in the data can then be used to train the models that can bring intelligence to the processes. This presents a challenge because a model trained on the data of historical actions will not only learn the fundamentals of the decision-making for the process involved, but it will also inherit the bias that the people running these processes had.

S. Agarwal, S. Mishra, *Responsible AI*,
https://doi.org/10.1007/978-3-030-76860-7_2

One of the most common practice in machine learning when solving problems that do not have sufficient historical data to train a model from scratch is to use a model trained for a similar problem and retrain it using the limited data available (also called transfer learning) or just train a model using the data from similar use cases. This can introduce new biases where the trained model has an unintended consequence of using the adjacent data.

The possible impact of the biases is not limited to specialized use cases. As AI adoption increases, there will hardly be any aspect of our lives that will remain untouched by this technology. It is not just the credit applications that utilize machine learning to determine whether an applicant gets loan or not. The applications now extend from machine learning algorithms evaluating medical test results for detecting illnesses to headphones that can customize their sound profile to a specific user based on their hearing. The impact of biases in this can potentially range from an annoyance at the headphone not sounding as good as you expected to a difference in life expectancy for a cancer patient based on the accuracy of the test result analysis. Some examples of such biases and the resulting discrimination that stand out from the recent news are the following:

- Amazon found that their machine learning-driven recruiting tool was showing bias against women applicants. They ended up scrapping the tool.
- Bank of America's Countrywide Financial business has agreed to pay a record fine of $335 m (£214 m) to settle discrimination charges when around 200,000 qualified African American and Hispanic borrowers were charged with higher rates solely because of their race or national origin.
- In 2019, Apple Card's algorithm gave lower credit limit to women as compared to men.
- Google's photo recognition AI led to coloured people being misidentified as primates.

An AI-driven product with biases against a section of users is not just bad news for the affected users anymore. It is equally bad for generating new business and has a significant reputational impact with possibility of fines from the regulators. The users and the regulators expect products to not just be fair to the users, but also be able to explain their actions reliably.

In this chapter, we will cover the key metrics for fairness, to be able to better understand the biases creeping into the model and to overcome the discrimination that they can cause. To be able to define fairness metrics and ways to measure them, let's introduce some key concepts.

The first concept we want to introduce is of favourable and unfavourable outcomes of the modelling problem – based on the problem you are trying to solve. The "favourable" outcome is the one that the user desires – for a home buyer, it will be approval for the mortgage application, whereas for a headphone buyer, it can be successful configuration of the wearable. An "unfavourable" outcome on the other hand is the one that the user does not desire – for a cancer patient getting a CT scan, it can be incorrect diagnosis, and for a smartphone user, it can be the voice assistant not recognizing their accent however hard they try. Throughout the book, we will be denoting the favourable predicted outcomes with $\widehat{Y_{fav}}$ and the unfavourable pre-dicted outcomes with $\widehat{Y_{unfav}}$ whereas Y_{fav} denotes actual favourable outcomes and Y_{unfav} denotes actual unfavourable outcomes.

The protected features and privileged/unprivileged classes are the next concepts we want to talk about. The features in your datasets can be divided into two groups – independent features and protected features. Independent features do not contain any personal, racial or socio-economic indicators that may be used for discrimination, whereas the protected features are the ones that may contain such information. For a given protected feature, we will have different user groups or classes, e.g. Marital Status can have the values Single, Married or Divorced. Based on the problem you are trying to solve, one of the groups within the Marital Status can be privileged with Divorced people receiving privileged treatment as compared to Single or Married. In this case, "Divorced" is the advantaged or privileged class, whereas Married or Single will be disadvantaged or unprivileged classes. Similarly, based on another protected feature, let's say Gender, there can be a different privileged class (e.g. Male) and unprivileged classes (Female, Other). Throughout the book, we will be using S to represent protected features with S_a denoting privileged class and S_d denoting unprivileged classes. The independent features will be represented using X.

The determination of favourable and unfavourable outcomes is based on the business problem being solved. For determining the protected features, we can start with a list of features that are widely known to introduce bias. Through the course of this book, we will cover how you can identify protected features given the scope of the problem you are trying to solve. The determination of privileged classes and unprivileged classes should always be done based on the data available for the problem. It is useful to plot heat maps of the frequency for each of the protected features by target outcome label and to select the class with the highest frequency of the favourable outcome. This has been demonstrated with an example later in this chapter. This is because a privileged class for one example may be unprivileged class for another. By transferring our understanding of privileged class from one problem to another, we may end up compounding the bias and not actually reduce it.

Here's an illustrative list of widely recognized protected features:

- Race (Civil Rights Act of 1964)
- Colour
- Sex including gender, pregnancy, sexual orientation, and gender identity (Equal Pay Act of 1963)
- Religion or creed
- National origin or ancestry
- Citizenship (Immigration Reform and Control Act)
- Age (Age Discrimination in Employment Act of 1967)
- Pregnancy (Pregnancy Discrimination Act)
- Familial status
- Physical or mental disability status (Rehabilitation Act of 1973)
- Veteran status
- Genetic information

Key Parameters

Before proceeding further with the chapter, let's recap the key parameters that we will use through the rest of this chapter to introduce more concepts and through the rest of the book as we go in more detail of identifying the existing biases and work on removing them.

– For any problem we are trying to solve, or any model we are working to train, the various possible outputs of the model are the possible outcomes – these outcomes can be favourable or unfavourable.
– Protected features can introduce biases in our models. We have looked at some examples of protected features recognized by law, but in order to remove the biases efficiently for our use case, we need to identify all protected features for our dataset.
– The possible user classes that comprise a given protected feature will have a privileged and unprivileged groups.
– Privileged groups are the ones that are more likely to have a favourable outcome than the unprivileged groups.

Any bias reduction approach will aim to reduce this gap. To calculate this gap and to define the fairness metrics, we will introduce some more concepts that are central to making these determinations.

Confusion Matrix

By definition, a confusion matrix M is such that $\sum_{i=1}^{n}\sum_{j=1}^{n} M$ is equal to the number of observations known to be in group Y_i and predicted to be in group $\widehat{Y}_j$.

The quadrants of the confusion matrix help explain:

1. **True positive (TP):** a case when the predicted and actual outcomes are both in the positive class
2. **False positive (FP):** a case when predicted values are in the positive class $(\widehat{Y_{fav}})$ when the actual outcome belongs to the negative class (Y_{unfav})
3. **False negative (FN):** a case when predicted values are in the negative class $(\widehat{Y_{unfav}})$ when the actual outcome belongs to the positive class (Y_{fav})
4. **True negative (TN):** a case when the predicted and actual outcomes are both in the negative class

These four parameters then help us define the accuracy metrics.

Common Accuracy Metrics

Figure 2.1 shows the commonly used accuracy metrics and their relationship with confusion matrix. Let's look at them in more detail.

- The **false positive rate (FPR)** is calculated as FP / (FP + TN) (where FP + TN is the total number of negatives). It's the probability that a false alarm will be raised. A positive result will be given when the true value is negative. This is something that a business mostly wants to minimize. The cost of false positive rate can be very expensive in case of financial services offerings.
- The **false negative rate (FNR)** – also called the miss rate – is the probability that a true positive will be missed by the test. It's calculated as FN / (FN + TP) (where FN + TP is the total number of positives). Again, this is something that may cost a business heavily.
- The **true positive rate (TPR)** (also called **sensitivity/recall**) is calculated as TP/ (TP + FN). TPR is the probability that an actual positive will test positive. Recall is the ability of the classifier to find all the positive samples. This is one of the most important secondary model metrics as it talks about increasing the favourable outcome.
- The **true negative rate (TNR)** (also called **specificity**), which is the probability that an actual negative will test negative. It is calculated as TN/(TN + FP). This is also important in certain cases especially when negative outcomes can be of high cost.

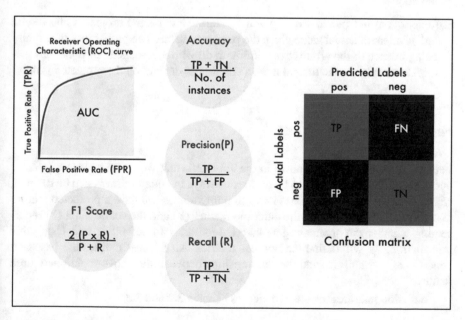

Fig. 2.1 Commonly used accuracy metrics and their relationship with confusion matrix

- The ratio of TPR + FNR and FPR + TNR will always equate to 1.
- **Positive predictive value (PPV)** (also known as **precision**): the fraction of positive cases correctly predicted to be in the positive class out of all predicted TP positive cases. Precision is the ability of the classifier not to label as positive a sample that is negative. TP/(TP + FP).

With the key parameters defined, let's bring in the definition for fairness and the fairness metrics.

Fairness and Fairness Metrics

As defined in fairness and machine learning by Barocas, Hardt and Narayanan, the concept of fairness revolves around majorly the below three concepts, that are based on the relationships between the protected feature and the actual and the predicted outcomes.

Independence: $\hat{Y}$ independent of S, if the predicted features and protected features are independent of each other. In other words, probability of being in a favourable (or non-favourable) class has nothing to do with the group of S.

Separation: $\hat{Y}$ independent of S given Y, if the predicted feature given the target value Y is independent of protected feature. The predicted probability of being in any class, given their actual class, has nothing to do with their protected group membership.

Sufficiency: Y independent of S given $\hat{Y}$. Here we expected protected class to be independent of actual value given the predicted value. The probability of actually being in each of the groups has nothing to do with membership of protected feature. In simple words, the prediction should not depend on the protected group.

Fairness Metrics

Let's consider an example, shown in the Table 2.1, that we can use to start defining our fairness metrics. There can be numerous ways in which fairness can be defined, we'll cover some important ones here. In our example, we have a protected feature (S) representing Gender and the outcome feature ($\hat{Y}$) and the actual label (Y) representing whether the applicant gets labelled as likely to default or not. Using this information, we will define the fairness metrics. All fairness metrics are based on conditional probabilities and the relationships between the above-mentioned three features.

Confusion matrices for the two classes (Tables 2.2 and 2.3)

Table 2.1 Sample actual and predicted likelihood of user defaulting on a loan and the gender of the user

Protected feature (Gender)	M	M	M	M	M	M	M	M	M	M	M	M	M	M	M	F	F	F	F	F	F
Actual	Y	Y	Y	Y	Y	Y	Y	Y	Y	Y	Y	N	N	N	N	Y	Y	Y	N	N	N
Predicted	Y	Y	Y	Y	Y	Y	Y	Y	N	N	N	Y	Y	N	N	Y	N	N	Y	N	N

Table 2.2 Confusion matrix for the protected group "Gender = Male"

Male		Actuals (Y)	
		P	N
Predicted ($\hat{Y}$)	P	8	2
	N	3	2

TP	FP	TN	FN
8	2	2	3

Table 2.3 Confusion matrix for the protected group "Gender = Female"

Female		Actuals (Y)	
		P	N
Predicted ($\hat{Y}$)	P	1	1
	N	2	2

TP	FP	TN	FN
1	1	2	2

Equal Opportunity

Both privileged and unprivileged groups have equal FNR.

$$P\left(\hat{Y} = 0 \,|\, Y = 1, S = S_a\right) = P\left(\hat{Y} = 0 \,|\, Y = 1, S = S_d\right)$$

A classifier satisfies this definition if both advantageous and disadvantageous groups have equal FNR. In the case of a loan dataset, this implies that the probability of an actual non-defaulting applicant to be incorrectly predicted as a defaulter should be the same for both groups of a protected feature of the protected feature applicants; no group will have an advantage of a reduced miss rate.

No one group should be penalized by allowing the model to make higher mistakes compared to the other group. Since $TPR + FNR = 1$, a classifier with equal FNRs will also have equal TPRs (also called sensitivity/recall). In our illustration in

Table 2.2, the FNR for Males is 0.273, whereas using Table 2.3, the FNR for Females is 0.667 – which demonstrates Women do not have equal opportunity of a favourable outcome in our dataset as compared to Men, which in turn establishes that, for this dataset, Women are the unprivileged class.

Predictive Equality

Both privileged and unprivileged groups have equal FPR.

$$P\left(\hat{Y} = 1 \mid Y = 0, S = S_a\right) = P\left(\hat{Y} = 1 \mid Y = 0, S = S_d\right)$$

A classifier satisfies this definition if both advantageous and disadvantageous groups have equal FPR. In our example, this implies that the probability that of an actual defaulter to be incorrectly predicted as a non-defaulter should be the same for both subsets of the protected class applicants.

In our illustration in Table 2.2, the FPR for Males is 0.5, whereas for Females it is 0.333, which means that Males are more likely to receive labelled as unlikely to default when they actually will default as compared to Women, confirming our finding above.

Equalized Odds

Also known as disparate mistreatment, equalized odds requires that both advantageous and disadvantageous groups have equal TPR and FPR.

$$P\left(\hat{Y} = 1 \mid Y = i, S = S_a\right) = P\left(\hat{Y} = 1 \mid Y = i, S = S_d\right), i \subset 0,1$$

This definition combines the previous two: a classifier satisfies the definition if advantageous and disadvantageous groups have equal TPR and equal FPR. In our example, this implies that the probability of a loan applicant with an actual good outcome (i.e. a non-defaulter) to be correctly predicted as a non-defaulter and the probability of a loan applicant with an actual bad outcome (i.e. a defaulter) to be incorrectly predicted as a non-defaulter should both be same for both sides of the protected group. The main idea behind the definition is that loan applicants with a good actual credit score and loan applicants with a bad actual credit score should have a similar classification, regardless of their membership to a protected class.

This shows how different metrics give different outcomes and the appropriate metric should be chosen depending on the use case or the principles of the use case.

Predictive Parity

Predictive parity is also known as the **outcome test**; this requires both advantageous and disadvantageous groups to have equal PPV/precision.

$$P\left(Y=1\mid \hat{Y}=1, S=S_a\right)=P\left(Y=1\mid \hat{Y}=1, S=S_d\right)$$

The main idea behind this definition is that the percentage of correct positive predictions should be the same for both the groups of a protected class. It implies that the errors are spread homogeneously among all the groups of a protected class. Furthermore, the odds for receiving a favourable outcome would be same irrespective of their membership to the group. In our example, this implies that, for both the advantageous and disadvantageous subsets of a protected group, the probability that a non-defaulting loan applicant is predicted to be a non-defaulter should be the same.

Predictive parity has one significant advantage. A perfectly predictive model is deemed as fair because both men and women would have a PPV of 1. A disadvantage of this model is that it doesn't necessarily reduce bias.

Demographic Parity

Membership in a protected class should have no correlation with being predicted a favourable outcome.

$$P\left(\hat{Y}=1, S=S_a\right)=P\left(\hat{Y}=1, S=S_d\right)$$

It suggests that the composition of the set of individuals who are predicted the favourable outcome should reflect the percentage of applicants. In our example, if 40% of loan applicants are female, then 40% of the applicants predicted to be non-defaulters must be female.

If we consider the confusion matrices of both privileged and unprivileged groups, we can see that to achieve demographic parity, for the privileged group, we either:

- For the privileged group, we reduce false positives and increase true negatives. This is the ideal case. Here not only the cost would decrease as FP is more expensive for the business but also ensuring that unfavourable outcomes are kept at a bay to prevent any gain by privileged group. This is akin to reducing the privilege of the privileged class.
- For the unprivileged group, we reduce false negatives and increase true positives. Here we will observe that the probability of unprivileged class to have a favourable outcome (increase in TP) will see a boost and will also ensure that unprivileged class does not see a decline in the favourable outcome (reduced FN) when they deserve a favourable outcome. This will also increase the number of candidates to whom service should be offered and is similar to positive discrimination.

Average Odds Difference

Average of difference in FPR and TPR for advantageous and disadvantageous groups.

$$\frac{1}{2}\left[\left(FPR_{s_d} - FPR_{s_a}\right) + \left(TPR_{s_d} - TPR_{s_a}\right)\right]$$

This metric encompasses both the predictive equality difference (FPR to TNR diff) and equal opportunity difference. A lower difference (or a zero difference) would mean equal benefit (positive or favourable outcome) for both the group.

Table 2.4 shows the discrimination based on fairness metrics.

In an ideal scenario, we would expect near identical values, if not the same, for all the fairness metrics across protected classes (i.e. the difference between them will be zero or near zero). However, for our example, we can see that the difference between the fairness metrics for Males and Females is significant in proportion to the absolute measure of these metrics for either of the class. From Table 2.4, it is evident that most of the metrics show the discrimination against the protected group "Gender = F" (or Females are an unprivileged group for our problem at hand).

You would need to compute these metrics for all protected features and all groups within those features to get a correct understanding of privileged vs unprivileged groups. Once the determination has been made, then based on the problem you are working to solve, you'd need to prioritize the metrics and would track them as you implement the different techniques to remove the bias covered in the book.

Note: Prioritizing Fairness Metrics

As you work on reducing the bias, the techniques you deploy will help you reduce the difference between the fairness metrics for the prioritized privileged and unprivileged classes; however, the improvement you achieve is not always the same across all fairness metrics and at times you may get significant improvement for some, whereas for other fairness metrics, the difference may actually increase. This makes it really important to prioritize the fairness metrics that we then aim to improve as part of bias reduction in our predictions. Let's see how different fairness metrics can conflict with each other to understand this better; we will continue to use our example.

Demographic Parity: Demographic parity is achieved when the probability of applicants from the advantaged group having favourable outcome is same as the probability of applicants from the disadvantaged group having favourable outcome. More often than not, the advantaged group overshadows disadvantaged groups in data volume. This means that any classifier trained will understand the advantaged group better and will be able to identify the applicants that are likely to default in the future much better than it can do for the disadvantaged classes – essentially a challenge due to less data available. In order to achieve demographic parity, it would then end up assigning favourable outcomes for the applicants that are highly likely to default in the future, and this will reduce the accuracy of the model.

Hence, demographic parity can be a good metric to improve on if we have sufficient volume of data for the advantaged as well disadvantaged groups. In absence of sufficient data, it is probably a good idea to utilize other metrics to improve the fairness.

Equal Opportunity: This requires that the records that get awarded favourable outcome meet or exceed the same qualifying criteria regardless of which protected group they belong to. Or simply put, equal opportunity takes a merit-based approach like you'd see in a selection of a sports team – well most of the time. This overcomes the issue that demographic parity presents but has challenges of its own.

Based on the problem you are trying to solve; equal opportunity can make disadvantaged class' representation in the group with favourable outcome sparse. This would not just impact other fairness metrics but more importantly can completely fail to address the reason for the poor performance on demographic parity. In Chicago, the mortgage lenders invested more in a single white neighbourhood than all the black neighbourhoods combined (ref: Where Banks Don't Lend). This disparity won't reduce if we apply equal opportunity because if the mortgage applicant is trying to buy a house in a black neighbourhood, the model is likely to give more weight to the address which has more nearby addresses with mortgages. Just the fact that the black neighbourhoods did not see as much investment in mortgages in the past would mean that the discrimination continues into the future.

In such a case, we need to find an approach that helps reduce the discrimination. That answer may lie in demographic parity – we may want to reduce the difference between the two groups while monitoring the impact on the accuracy and keeping that within acceptable limits.

Table 2.4 Fairness metrics for the two protected groups

Metrics	Male	Female	Difference
Statistical parity	0.733	0.5	0.233
Demographic parity	0.667	0.3	0.333
True positive rate	0.727	0.3	0.427
True negative rate	0.500	0.7	−0.167
False positive rate	0.500	0.333	0.167
Equal opportunity/false negative rate	0.273	0.667	−0.394
PPV/predictive parity	0.800	0.5	0.300
Equalized odds	1.227	0.666	0.561
Average odds ratio			−0.280

Python Implementation

The code below shows how to compute the various metrics that we have discussed so far.

```
def fair_metrics(y_actual, y_pred_prob, y_pred_binary, X_test, protected_group_name,
                 adv_val, disadv_val):
    """
    Fairness performance metrics for a model to compare advantageous and disadvantageous groups of a pr
otected variable

    Parameters
    ----------

    :param y_actual: Actual binary outcome
    :param y_pred_prob: predicted probabilities
    :param y_pred_binary: predicted binary outcome
    :param X_test: Xtest data
    :param protected_group_name: Sensitive feature
    :param adv_val: Priviliged value of protected label
    :param disadv_val: Unpriviliged value of protected label
    :return: roc, avg precision, Eq of Opportunity, Equalised Odds, Precision/Predictive Parity, Demogr
aphic Parity, Avg Odds Diff,
             Predictive Equality, Treatment Equality

    Examples
    --------
    fairness_metrics=[fair_metrics(y_test, y_pred_prob, y_pred,
                      X_test, choice, adv_val, disadv_val)]


    """
    tn_adv, fp_adv, fn_adv, tp_adv = confusion_matrix(
        y_actual[X_test[protected_group_name] == adv_val],
        y_pred_binary[X_test[protected_group_name] == adv_val]).ravel()

    tn_disadv, fp_disadv, fn_disadv, tp_disadv = confusion_matrix(
        y_actual[X_test[protected_group_name] == disadv_val],
        y_pred_binary[X_test[protected_group_name] == disadv_val]).ravel()

    # Receiver operating characteristic
    roc_adv = roc_auc_score(y_actual[X_test[protected_group_name] == adv_val],
                            y_pred_prob[X_test[protected_group_name] == adv_val])
    roc_disadv = roc_auc_score(y_actual[X_test[protected_group_name] == disadv_val],
                               y_pred_prob[X_test[protected_group_name] == disadv_val])

    roc_diff = abs(roc_disadv - roc_adv)

    # Average precision score
    ps_adv = average_precision_score(
        y_actual[X_test[protected_group_name] == adv_val],
        y_pred_prob[X_test[protected_group_name] == adv_val])
    ps_disadv = average_precision_score(
        y_actual[X_test[protected_group_name] == disadv_val],
        y_pred_prob[X_test[protected_group_name] == disadv_val])

    ps_diff = abs(ps_disadv - ps_adv)
```

```
# Equal Opportunity - advantageous and disadvantageous groups have equal FNR
FNR_adv = (fn_adv)/(fn_adv + tp_adv)
FNR_disadv = (fn_disadv)/(fn_disadv + tp_disadv)
EOpp_diff = abs(FNR_disadv - FNR_adv)

# Predictive equality  - advantageous and disadvantageous groups have equal FPR
FPR_adv = (fp_adv)/(fp_adv + tn_adv)
FPR_disadv = (fp_disadv)/(fp_disadv + tn_disadv)
pred_eq_diff = abs(FPR_disadv - FPR_adv)

# Equalised Odds - advantageous and disadvantageous groups have equal TPR + FPR
TPR_adv = (tp_adv)/(tp_adv + fn_adv)
TPR_disadv = (tp_disadv)/(tp_disadv + fn_disadv)
EOdds_diff = abs((TPR_disadv + FPR_disadv) - (TPR_adv + FPR_adv))

# Predictive Parity - advantageous and disadvantageous groups have equal PPV/Precision (TP/TP+FP)
prec_adv = (tp_adv)/(tp_adv + fp_adv)
prec_disadv = (tp_disadv)/(tp_disadv + fp_disadv)
prec_diff = abs(prec_disadv - prec_adv)

# Demographic Parity - ratio of (instances with favourable prediction) / (total instances)
demo_parity_adv = (tp_adv + fp_adv) / (tn_adv + fp_adv + fn_adv + tp_adv)
demo_parity_disadv = (tp_disadv + fp_disadv) / (tn_disadv + fp_disadv + fn_disadv + tp_disadv)
demo_parity_diff = abs(demo_parity_disadv - demo_parity_adv)

# Average of Difference in FPR and TPR for advantageous and disadvantageous groups
AOD = 0.5*((FPR_disadv - FPR_adv) + (TPR_disadv - TPR_adv))

# Treatment Equality  - advantageous and disadvantageous groups have equal ratio of FN/FP
TE_adv = fn_adv/fp_adv
TE_disadv = fn_disadv/fp_disadv
TE_diff = abs(TE_disadv - TE_adv)

return [('AUC',roc_diff),('Avg PrecScore', ps_diff),('Equal Opps', EOpp_diff),
        ('PredEq', pred_eq_diff), ('Equal Odds', EOdds_diff), ('PredParity', prec_diff),
        ('DemoParity', demo_parity_diff), ('AOD', abs(AOD)), ('TEq', TE_diff)]
```

Proxy Features

So far, we have focussed on the protected features for determining if our data has any bias and for defining and calculating fairness metrics. We defined independent features as without any information that may be used for information. That is true in an ideal scenario, but it is far too common for any reasonably complex problem to continue to have bias in the data even after removing the protected features. It's a common mistake to assume that if you are not using any protected features, then you cannot have bias in your dataset and that it makes your model immune from making discriminatory decisions.

This is because often we have independent features in our dataset that represent a protected feature even though you may not have that feature in your data anymore. In other words, these independent features serve as a proxy feature for the protected features.

The relationship between proxy features and the protected features can be explained most of the times using multicollinearity. Multicollinearity is defined as highly linear relationship between more than two variables. If the correlation between two independent variables is 1 or -1, we have perfect multicollinearity between them. We will introduce few mechanisms to check for multicollinearity; before that, let's look at some examples of proxy features that are commonly used:

- Tax paid is a reflection of income.
- Zip code can be used to derive race/ethnicity.
- Shopping pattern can talk about gender and marital status.
- Disposal income can reveal gender and marital status.
- Salary and age can reveal gender and promotions.

It is important to realize and acknowledge that the presence of such proxies can lead to discrimination just because, in simple terms, they are correlated to protected features in some form. The extensive use of (undetected) proxies in model development has raised concerns on fairness. Not only is it tough to detect but also tough to process the information by a human. Fairness definition and measurement is futile if one misses on detection of proxy feature. In such a case, despite all measures, discrimination would still be present in the decision-making.

The successful determination of proxy features in the dataset is an important step in the process to remove bias. In addition to the methods that can help the decision, the business context and the problem statement being solved are also key to making the determination. As a result, the business analysts and the product owner have a very important role to play along with the data scientists in the team. Sometimes, a proxy feature gets formed after feature engineering, for instance, disposable income – which is dependent on the total income and the month-end balance. This engineered feature is seen to be proxy of gender or marital status. Disposable income is seen to be high in case of single male compared to married male. So even if gender or marital status is removed from the data, this one feature would still contain that information and can introduce biases in the prediction. Hence, it's very important to detect proxies at various stages including pre- and post-feature engineering stages.

Catherine O'Neil's Weapons of Math Destruction cites an example where an individual's browsing history and location was used as proxies for determining ability to repay the loans which helped in deciding if ads for payday loans should be displayed to that individual.

In another example, in 2019, American Express send letters to customers informing them that their credit limit is being decreased because "Other customers who have used their card at establishments where you recently shopped have a poor repayment history with American Express".

Consider the following example: You created a credit decision model to make decisions on auto loan applications. You have taken care to remove all personal and racial identifiers from your dataset. Yet, when you test your model, you find that it is favouring men over women. You know that the original data showed bias in favour of men, but since you removed all protected fields from the dataset, you expected it to drop the bias. What can possibly be the reason for the unresolved bias?

When you analyse the data to find if there are any proxy protected features, you realize that the car engine size serves as a proxy for buyer's gender! A proxy like this can go undetected just because the proxy feature is considered to be part of the core data that is needed for solving the problem. There are many mechanisms available to us for handling this as we will see in the following chapters.

Undetected proxies for protected features not only undermine the efforts to reduce bias from the data and the predictions; they also reduce the ability to explain the output. Identifying protected features and their relationship with other features to find any proxies should be one of the first activities in your model's lifecycle. However, there are instances where you have models in production for which you need to address the bias. Handling them is covered in more detail in a later chapter on Reject Option Classifiers, but even then, it is useful to identify the proxy features in your dataset, if any, before applying the reject option technique to reduce bias.

Methods to Detect Proxy Features

Linear Regression

In order to detect proxy features, let's begin with a very simple method. For instance, we have only two features, and both are related or co-related. In such case, if we develop a linear regression model with one being the independent feature and other being a dependent feature, the results will reveal a relationship. The equation below is a relationship between X1 and X2.

$$X_1 = a + \lambda * X_2$$

The standard error in regression is given by

$$SE = \sqrt{1 - R^2_{adj}} \times \sigma_y$$

$$\text{where } R^2_{adj} = 1 - \frac{\left(1 - R^2\right)\left(n - 1\right)}{n - k - 1}$$

Here Y_i is the actual value, $\widehat{Y}_i$ is the prediction, X_i is the independent feature, $\overline{X}_i$ is the mean, n is no of rows and k is number of features. R^2 is the R-square value of the regression representing closeness of the actual data to the fitted regression line. In case of multicollinearity for the above discussed case, (if the two features are identical) $R^2 = 1$ would represent a perfect fit model, and thus the standard error will be 0. Using this as a preliminary check, a regression test should be developed for protected features as dependent features with all other features. However, this approach is also very computation intensive, making it difficult to shortlist all the proxies if the feature space is large.

Variance Inflation Factor (VIF)

Taking the above concept forward, one of the widely used solution in multicollinearity is VIF. VIF looks at the coefficient of determination (R^2), for each explanatory variable. This method is widely used to remove collinear or multicollinear features during any statistical modelling. This method can also be used to detect and remove features that are proxies to a given protected/sensitive features. Typically, this is used for dropping multicollinear (redundant) features. VIF internally develops multiple regression among all features and returns the list of features that are correlated.

$$VIF = \frac{1}{1 - R^2}$$

A high value for VIF indicates collinearity (collinearity results in R^2 value close to 1, which leads to a high VIF). Conventionally a feature is said to be collinear with any other feature if VIF is more than 5 (or 7 depending on who you ask). But the threshold can be altered looking at sensitivity of protected features and appetite of the business. A high value of VIF in a regression between a protected features and other features would mean a strong presence of another feature that is collinear and thus is a proxy to the feature under investigation.

For instance, we created a dummy data where Feature 1 to Feature 4 are related with each other and Feature 5 is a random feature and has little/nothing to do with the first four features. We aim to find proxy(ies) for Feature 1 as it's one of the protected features that can eventually result in discrimination.

	Feature 1	Feature 2	Feature 3	Feature 4	Feature 5
0	34	34	1156	6.8	8
1	71	71	5041	71.0	8
2	88	88	7744	88.0	8
3	85	85	7225	85.0	11
4	35	35	1225	35.0	9

To start with, looking at the correlation matrix, it would be evident how the first four features seem to be strongly related to each other. The first four features (Fig. 2.2) have a strong relation with Feature 1 and are also related with each other. However, Feature 5 shows a weak relationship with the first four features.

	Feature 1	Feature 2	Feature 3	Feature 4	Feature 5
Feature 1	1.000000	1.000000	0.975059	0.990414	-0.078924
Feature 2	1.000000	1.000000	0.975059	0.990414	-0.078924
Feature 3	0.975059	0.975059	1.000000	0.967815	-0.122876
Feature 4	0.990414	0.990414	0.967815	1.000000	-0.065173
Feature 5	-0.078924	-0.078924	-0.122876	-0.065173	1.000000

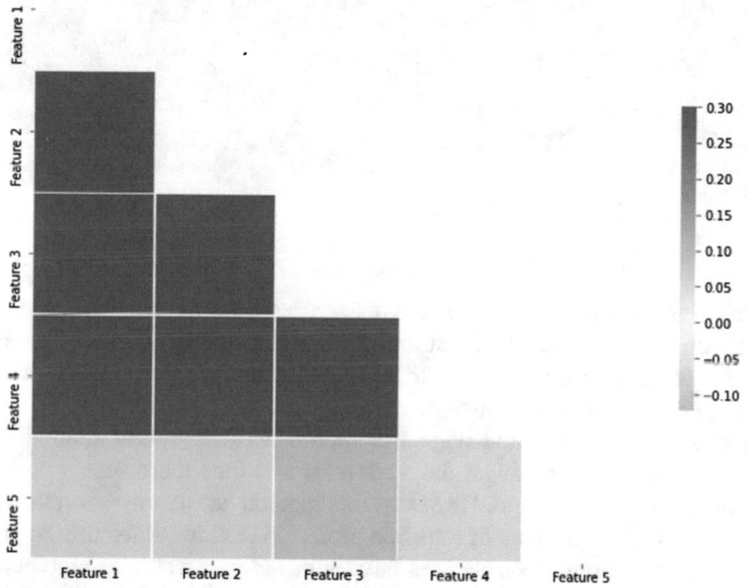

Fig. 2.2 Heat map showing Pearson correlation between the five features

In order to better understand the relationship, we attempted a univariate linear regression between Feature 1 (as the dependent Y variable) and other feature (as independent X) one at a time. This is a crude method of finding any linear relationship which can be further used to find VIF values.

```
for i in np.arange(1,5):
    y=proxy['Feature 1']
    X=proxy.iloc[:,i]

    res = stats.linregress(X,y)
    r2=res.rvalue**2
    print(f"R-squared: {res.rvalue**2:.6f}")
    vif = 1/(1-r2)
    print ('VIF of Feature {} with {}'.format(i+1, "Feature 1"), vif)
    print("")
R-squared: 1.000000
VIF of Feature 2 with Feature 1 inf

R-squared: 0.950740
VIF of Feature 3 with Feature 1 20.30056335221401

R-squared: 0.980920
VIF of Feature 4 with Feature 1 52.40998864414569

R-squared: 0.006229
VIF of Feature 5 with Feature 1 1.006268068628911
```

It can be seen that VIF values of Features 1 with the other features except Feature 5 is quite high, showing strong multicollinearity between each combination. Furthermore, we can use VIF to drop features that are related, and we would see a similar outcome.

```
def calculate_vif_(X, thresh=5.0):
    variables = [X.columns[i] for i in range(X.shape[1])]
    dropped=True
    while dropped:
        dropped=False

        vif = Parallel(n_jobs=-1,verbose=5)(delayed(variance_inflation_factor)(X[variables].values, ix)
                                            for ix in range(len(variables)))

        maxloc = vif.index(max(vif))
        if max(vif) > thresh:
            print ('VIF is ' + str(max(vif)) + ' ' + 'dropping ' + X[variables].columns[maxloc])

            variables.pop(maxloc)
            dropped=True

    print('Remaining variables:')
    print([variables])
    return
```

```
calculate_vif_(proxy)
```

```
VIF is inf dropping Feature 1
VIF is 299.3857233076956 dropping Feature 2
VIF is 70.87650502990464 dropping Feature 4
Remaining variables:
[['Feature 3', 'Feature 5']]
```

Here we are recursively dropping the feature that has maximum VIF, and then we are left with only those features that are independent and not related. This process dropped Feature 1, Feature 2 and Feature 4 and left us with two non-correlated features, viz. Feature 3 and Feature 5.

VIF comes with its own advantage. For starters, we can say that multicollinearity is merely a correlation amongst the variables, and thus there seems no harm to assess the pairwise correlation. However, this method has its own shortcoming too. Imagine if a proxy is made up of multiple protected feature (or features in general), in such case the pairwise correlations between each pair won't be high enough to raise an alarm. But this won't mean that collinearity is absent. In such a scenario, a concept like VIF may come to your rescue.

Linear Association Method Using Variance

Finally, we will cover one more method of determining collinearity. The method proposed in the paper (Hunting for Discriminatory Proxies in Linear Regression Models) talks about computing association between two variables. This uses square of Pearson correlation coefficient in order to highlight a higher association measure to represent a proxy with a higher value.

$$Assoc = \frac{cov(X_1, X_2)^2}{Var(X_1)Var(X_2)}$$

Using this method on the above dummy dataset, we would see similar results as above.

```
for i in np.arange(1,5):

    var1 = np.var(proxy['Feature 1'])
    var2 = np.var(proxy.iloc[:,i])
    cov = np.cov(proxy['Feature 1'], proxy.iloc[:,i])[0][1]

    asso = np.square(cov) / (var1*var2)

    print ('Association of Feature {} with {} :'.format(i+1, "Feature 1"), asso)
```

```
Association of Feature 2 with Feature 1 : 1.041232819658476
Association of Feature 3 with Feature 1 : 0.989941985921839
Association of Feature 4 with Feature 1 : 1.0213657514412269
Association of Feature 5 with Feature 1 : 0.006485864925820933
```

As expected, the association between Feature 1 and the other three features is very strong, showing presence of proxies. However, Feature 5 shows very low association with the first feature, thus inferring no proxy and no relationship.

Cosine Similarity/Distance Method

The relationships between the features may not always be linear, and in this case the above methods won't be able to tell us that any two features are related. A method that determines similarity (or distance) can help us determine if a given pair of features is related. One such method is cosine similarity, but you can choose to use other distance methods like Euclidean, Hamming, Jacquard, etc. depending on data type. This method can be used even for categorical proxies which can be really helpful in lot of cases as we will see eventually in the next few chapters.

Cosine similarity measures the cosine of the angle between two vectors to establish if they are pointing in the same direction or not. Cosine similarity helps finding similar items in multidimensional space and works based on the direction the vectors are pointing in. For example, cosine similarity can help us determine if two features (monthly income and retail expenditure) together can be a proxy for a loan applicant's Gender and no. of dependents together.

$$Similarity = \frac{A.B}{\|A\|\,\|B\|} = \frac{\sum A_i B_i}{\sqrt{\sum A_i^2}\,\sqrt{\sum B_i^2}}$$

Mathematically it's a dot product of two vectors divided by multiplication of Euclidean norm of the two vectors. On the one hand, where correlation returns the variability (how one value changes with respect to another), cosine similarity measures the similarity in the value itself. You can also find cosine distance between multiple vectors using the same above technique.

```
from scipy import spatial

for i in np.arange(1,5):

    result = 1 - spatial.distance.cosine(proxy['Feature 1'], proxy.iloc[:,i])
    print ('cosine distance between Feature 1 and Feature {}'.format(i+1), result)

    r = 1
    d = 10 * r * (1 - result)

    circle1=plt.Circle((0, 0), r, alpha=.5)
    circle2=plt.Circle((d, 0), r, alpha=.5)

    plt.ylim([-1.1, 1.1])
    plt.xlim([-1.1, 1.1 + d])
    fig = plt.gcf()
    fig.gca().add_artist(circle1)
    fig.gca().add_artist(circle2)
```

cosine distance between Feature 1 and Feature 2 1.0,
cosine distance between Feature 1 and Feature 3 0.96490,
cosine distance between Feature 1 and Feature 4 0.99787,
cosine distance between Feature 1 and Feature 5 0.87305.

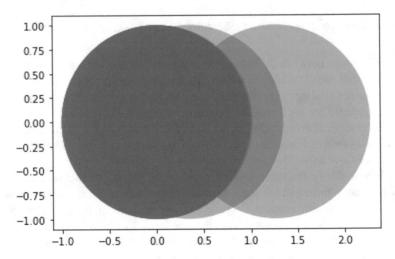

Fig. 2.3 Plot showing cosine similarity between the features

Here again, it is evident that cosine similarity between Feature 1 and the other three features are quite close and the distance between Feature 1 and Feature 5 is not so close. Even the graph (Fig. 2.3) shows how Feature 2 circle has got superimposed with Feature 1 and other feature 5 circle is at distinct distance from other features inferring less similarity to the first four.

Mutual Information

Mutual information measures the amount of information one can obtain from one random variable given another. It is often used as a general form of correlation coefficient. Given that a majority of time-protected features would be binary or categorical, we will be covering the topic for the discrete features. Let's consider two discrete random variables X_1 and X_2 with probability mass functions $p(x_1)$ and $p(x_2)$. Their covariance and mutual information values can be computed from below given formulas.

$$COV(X_1,X_2) = \Sigma \left[p(x,x_2) - p(x_1)p(x_2) \right] x_1 x_2$$

$$I(X_1,X_2) = \Sigma p(x_1,x_2) \left[\log p(x_1,x_2) - \log p(x_1)p(x_2) \right]$$

In both the cases, covariance and mutual information, we are computing distance between two given features. In case of $COV(X_1, X_2)$, a weighted sum of the product of the two features are created while in case of $I(X_1, X_2)$ a weighted sum of joint probabilities is computed. So, while one looks at impact on product, the other looks at impact on distribution. Thus, both speak same thing but in different diction. The plus point with mutual information is that it is not concerned about the way it is related. Covariance is concerned about only linear relationship, while MI don't care about linearity, which is why MI is also used for nonlinear tree based algorithms.

Here for illustration, we are using a set of categorical features where the first two categorical features are not related; Cat 3 is derived using logical operator on Cat 1 and Cat 2, whereas Cat 4 is an inverse of categorical feature Cat 1.

	Cat 1	Cat 2	Cat 3	Cat 4
0	0	0	0	1
1	1	0	1	0
2	2	1	0	0
3	3	1	1	0
4	4	1	1	0

```
from sklearn.metrics import mutual_info_score
for i in np.arange(6,9):
    mi = mutual_info_score(proxy['Cat 1'], proxy.iloc[:,i])
    print ('Mutual info between Cat 1 and {}'.format(proxy.iloc[:,i].name), mi)
```

```
Mutual info between Cat 1 and Cat 2 0.0014418857149821607
Mutual info between Cat 1 and Cat 3 0.19127284867825473
Mutual info between Cat 1 and Cat 4 0.6802920001921533
```

The low mutual information score between Cat 1 and Cat 2 is because of no relationship between the two features. The relationship between Cat 1 and Cat 4 is identified much more strongly, but the score for Cat 1 and Cat 3 indicates a possible relationship too. It is important to note that our sample data here is very small (only 5 records), and for a meaningful determination, we want the dataset to be a good representation of what our model will encounter in a production environment.

Conclusion

Calculating various fairness metrics and applying different techniques to reduce the bias across stages of the lifecycle for the model is not sufficient if proxy features are ignored. The various techniques covered in this chapter can be used to identify the proxies to help you decide the approach as you move ahead towards removing the bias. A lot of time a continuous feature can be a strong proxy of a binary protected feature. For instance, disposable income (a continuous data type) may be proxy for gender and/or marital status (a binary data type). Thus, it is imperative to spend enough time on finding direct proxies, linear proxies, nonlinear proxies and also combination of features which together can be proxy of one protected feature. Admittedly, this particular work is computationally heavy and would require exploratory data analysis. But then, compromising on this step would mean going back to square one despite implementing all fairness checks and balances.

Bibliography

Barocas, S., Hardt, M. and Narayanan, A., 2021. Fairness and machine learning. [online] Fairmlbook.org. Available at: <https://fairmlbook.org/> [Accessed 16 April 2021].

BBC News. 2021. Apple's 'sexist' credit card investigated by US regulator. [online] Available at: <https://www.bbc.co.uk/news/business-50365609> [Accessed 16 April 2021].

BBC News. 2021. Bank of America fined $335m for minority discrimination. [online] Available at: <https://www.bbc.co.uk/news/business-16296146> [Accessed 16 April 2021].

Brosnan, S., 2006. Nonhuman Species' Reactions to Inequity and their Implications for Fairness. Social Justice Research, 19(2), pp.153-185.

Dastin, J., 2021. Amazon scraps secret AI recruiting tool that showed bias against women. [online] U.S. Available at: <https://www.reuters.com/article/us-amazon-com-jobs-automation-insight-idUSKCN1MK08G> [Accessed 16 April 2021].

Hardt, M., Price, E. and Srebro, N. (2016). Equality of Opportunity in Supervised Learning. [online] . Available at: https://arxiv.org/pdf/1610.02413.pdf [Accessed 16 Apr. 2021].

Kanngiesser, P. and Warneken, F., 2012. Young Children Consider Merit when Sharing Resources with Others. PLoS ONE, 7(8), p.e43979.

Linda Lutton, A., 2021. Home Loans in Chicago: One Dollar To White Neighborhoods, 12 Cents To Black. [online] Interactive.wbez.org. Available at: <https://interactive.wbez.org/2020/banking/disparity/> [Accessed 16 April 2021].

Mrtz.org. 2021. nips17tutorial. [online] Available at: <https://mrtz.org/nips17/#/> [Accessed 16 April 2021].

Nytimes.com. 2021. American Express Kept a (Very) Watchful Eye on Charges (Published 2009). [online] Available at: <https://www.nytimes.com/2009/01/31/your-money/credit-and-debit-cards/31money.html> [Accessed 16 April 2021].

O'Neil, C., 2017. Weapons of math destruction. Great Britain: Penguin Books.

Pair-code.github.io. 2021. Playing with AI Fairness. [online] Available at: <https://pair-code.github.io/what-if-tool/ai-fairness.html> [Accessed 16 April 2021].

Sahil Verma and Julia Rubin. 2018. Fairness definitions explained. In Proceedings of the International Workshop on Software Fairness (FairWare '18). Association for Computing Machinery, New York, NY, USA, 1–7.

Simonite, T., 2021. When It Comes to Gorillas, Google Photos Remains Blind. [online] Wired. Available at: <https://www.wired.com/story/when-it-comes-to-gorillas-google-photos-remains-blind/> [Accessed 16 April 2021].

Uk.practicallaw.thomsonreuters.com. 2021. [online] Available at: <https://uk.practicallaw.thomsonreuters.com/5-501-5857?transitionType=Default&contextData=%28sc.Default%29> [Accessed 16 April 2021].

Yeom, S., Datta, A. and Fredrikson, M. (n.d.). Hunting for Discriminatory Proxies in Linear Regression Models. [online] . Available at: https://proceedings.neurips.cc/paper/2018/file/6cd9313ed34ef58bad3fdd504355e72c-Paper.pdf [Accessed 16 Apr. 2021].

Chapter 3
Bias in Data

Introduction

In this chapter, we are going to discuss how unintended bias can get introduced in the models through the data and how the product owner and the SMEs can work with the data scientists at the definition stage to identify the biases that can get introduced and make decisions on how to reduce or eliminate them.

Let's take a moment to talk about why data can have biases, where do these biases come from and what can be the impact of these biases if they are not addressed. Any machine learning model trained by the data scientists will require training data – which will come from the historical actions taken by the systems and the users.

For example, our sample loan dataset has 61,321 records with 205 features (columns) that describe each of these loans. The prediction model can predict two types of outcomes – categorical (yes/no, true/false classes) or numerical (scores for default). As we look at different mechanisms of determining bias, we will use this dataset for creating the relevant examples. For all the approaches, the output Y is the likelihood of the user defaulting on the loan, except when stated otherwise.

This list of features has been achieved after feature engineering, aggregation and one hot encoding (converting categorical variables into binary). As we will see soon that in this dataset, the likelihood of default on a loan for a debtor with three or more dependants is higher than the likelihood of default for a debtor with less than three dependants. If you are creating a feature to predict the likelihood of a debtor defaulting on their loan with this data, it will predict unfavourably against the people with three or more dependants. This can't have just adverse impact on the users but an adverse impact on the business and create a reputational/regulatory issue as well.

As part of defining the problem statement, it is important to identify the features that introduce the biases directly or indirectly (proxy features for biases). For the purpose of this chapter, we have identified around 15 features (converted to binary)

© The Author(s), under exclusive license to Springer Nature Switzerland AG 2021
S. Agarwal, S. Mishra, *Responsible AI*,
https://doi.org/10.1007/978-3-030-76860-7_3

that are tagged as protected or sensitive features (S). These are the features for which we expect disparity based on the value (e.g. for Gender, Male debtors may be favoured over Female debtors) leading to a privileged vs unprivileged class.

The sensitive features identified are Individual's Gender, Education, Age group, Home Ownership Type (Owner, Mortgages, Tenant), Employment Status, Language (Estonian, English, other), No. of Dependants (less than three or otherwise), Marital Status (Married, Single, Divorced, otherwise), Work Experience (less than ten, less than five or otherwise).

Thus, our 205 features are divided into 189 features (X_j), 15 sensitive features (S_j) and a binary labelled class (Y – identifying defaulters) with value 0,1. Out of the total 61,231 records, the data has about 76% non-defaulter instances (46,752) and 24% defaulter instances (14,569).

For each feature in the sensitive feature list (S_j), we will be using expressions S_a & S_d to denote the advantaged and disadvantaged groups, respectively. S_a is advantaged group of a protected attribute (S) that has shown higher probability of having a favourable outcome (Y_{fav}), while S_d is the disadvantaged group showing a lower probability of having a favourable outcome (Y_{fav}). In order to ensure the data to be fair, the probability of all groups (S_a & S_d) under protected class (S) needs to have same probability of favourable outcome (Y_{fav}).

To mathematically check for the difference in probability of having favourable outcome in the given dataset D, we will use the following two metrics: statistical parity difference and disparate impact.

Statistical Parity Difference

It is the difference between the probability that a random individual drawn from the disadvantaged group is labelled 1 and the probability that a random individual from the advantaged group is labelled 1. A bigger value indicates higher level of disadvantage.

It measures the difference that the majority and protected classes get a particular outcome. When that difference is small, the classifier is said to have "statistical parity", i.e. to conform to this notion of fairness. In simple words, statistical parity tells us if the protected class is statistically treated the same way as the general population. So, if 10% of Male people get loans, statistical parity requires roughly 10% of females to also get loans. Ideally, the two likelihoods should be same, and statistical parity difference should be 0:

$$SPD = P\left(Y = 1 \mid S = S_a\right) - P\left(Y = 1 \mid S = S_d\right) = 0$$

In a real-world scenario, however, you'd want to identify an upper limit for SPD for your sensitive features. This threshold does not need to be same for all the sensitive features, but that is a good place to start. For our example, we have set the SPD threshold for all sensitive features to be 0.1 (the acceptable values of SPD will range between 0 and 0.1).

Disparate Impact

Disparate impact is another measure of the adverse impact on the disadvantaged group as revealed by a feature. The Technical Advisory Committee on Testing, assembled by the State of California Fair Employment Practice Commission (FEPC) in 1971, came up with the 80% rule. Mathematically evaluated, this rule requires

$$\frac{P\left(Y=1\mid S=S_d\right)}{P\left(Y=1\mid S=S_a\right)} \geq 0.8$$

Or turning it around

$$\frac{P\left(Y=1\mid S=S_a\right)}{P\left(Y=1\mid S=S_d\right)} \leq 1.25$$

This requires the likelihood that a person in the disadvantaged group will have the positive outcome to be at least 80% of the likelihood that a person in the advantaged group will have the positive outcome.

For our example, we have set the threshold at 90% (higher is better). This means that we want the following condition to be fulfilled for all sensitive features (we rounded down 1/0.9 to 1.1 for simplicity).

$$\frac{P\left(Y=1\mid S=S_a\right)}{P\left(Y=1\mid S=S_d\right)} \leq 1.10$$

This would mean that the acceptable values of disparate impact for our example will be 1 to 1.1. In an ideal scenario, the disparate impact should be 1. It was observed that out of 15 protected features (S), around 5 protected features (No. of Dependants less than three, Married, Single, Divorced and Work Experience less than ten) are having bias as per statistical parity difference and disparate impact as shown in Fig. 3.1.

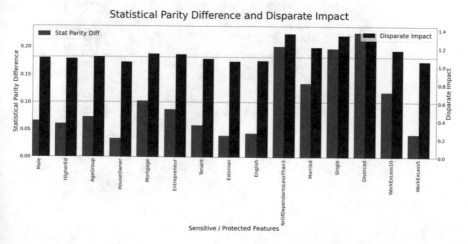

Fig. 3.1 Statistical parity difference and disparate impact for sensitive features

```
def statistical_parity_test(data_frame, protected_group, Sa_label, Sd_label,
                            Y, fav_label):

    '''

    Statistical Parity difference and disparate impact

    Parameters
    ----------

    :param data_frame: data frame with Y and S
    :param protected_group: Column name of protected feature
    :param Sa_label: binary label for advantageous group
    :param Sd_label: binary label for disadvantageous group
    :param Y: the actual label
    :param fav_label: binary label for favourable group
    :return: statistical_parity, disparate_impact

    Examples
    --------
    data = df
    protected_group = 'gender'
    Sa_label = 0
    Sd_label =1
    Y = 'Default'
    fav = 0
    statistical_parity, disparate_impact = statistical_parity_test(data, protected_group,
            Sa_label, Sd_label, 'Default', fav)


    '''

    Sa = data[data[protected_group] == Sa_label]
    Fav_Sa = Sa[Sa[Y] == fav_label]
    Fav_Sa_count = len(Fav_Sa)

    Sd = data[data[protected_group] == Sd_label]
    Fav_Sd = Sd[Sd[Y] == fav_label]
    Fav_Sd_count = len(Fav_Sd)

    Advantageous = len(Sa)
    dis_Advantageous = len(Sd)

    statistical_parity = (Fav_Sd_count / dis_Advantageous) - (Fav_Sa_count / Advantageous)
    disparate_impact = (Fav_Sd_count / dis_Advantageous) / (Fav_Sa_count / Advantageous)

    return statistical_parity, disparate_impact
```

It's clearly visible that five sensitive features (No. of Dependants less than three, Married, Single, Divorced and Work Experience Less than ten) meet the criteria of having bias as per statistical parity difference and disparate impact. However, it gets really tricky to handle multiple features that cause parity issues. It would also be interesting to see a Boolean condition on a few features that's seems to defy statistical parity.

Illustration:

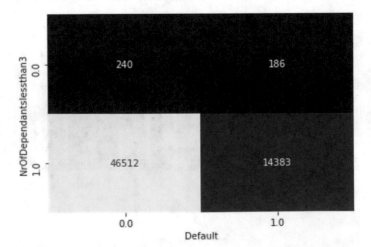

Fig. 3.2 Heat map for "No. of Dependants less than three"

SPD 0.2004
Disparate Impact 1.3558

(unprivileged count positive outcome / total unprivileged) - (privileged count positive outcome / total privileged)

Using the heat map as shown in Fig. 3.2:

SPD = |((240 / (240 + 186)) - (46,512 / (46,512 + 14,383)))| = |0.56338028 - 0.76380655| = 0.2004.
DI = 0.76380655 / 0.56338028 = 1.3558.

Similarly, we can calculate the SPD and DI for other sensitive features as shown in Table 3.1.

Table 3.1 SPD and DI values for various protected features

Heat map	SPD	DI
	0.0573	1.8003
	0.2258	1.3685
	0.1181	1.1762

Heat map	SPD	DI
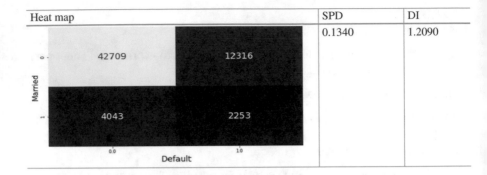	0.1340	1.2090

Now that we have identified the sensitive features, based on the number of features identified and the problem statement being solved, the product owner and the business analysts have a decision to make. If your findings are strong enough, you may want to move directly to the next stage and work on removing the bias.

Else, in the following sections in this chapter, we present some more techniques that you can utilize based on the feature type to explore the data further and identify the highest-priority sensitive feature set for your problem.

When the Y Is Continuous and S Is Binary

From our loan dataset, credit score can be an example of a continuous label Y. A corresponding protected feature can be the binary feature "married", which indicates if the person is married at the time of the record being captured or not.

In addition to the technique so far, for such a case, it would be interesting to draw a density plot of the protected feature, see Fig. 3.3, and note the difference in the distribution among the target value for each of the protected groups.

We will use three measures to compare the distributions of advantaged and disadvantaged classes.

Mean of the distribution: In this case, it will be the average credit scores for the two classes. Ideally, we want the difference in the mean values of the two distributions to be as close to 0 as possible.

Skewness: It is a measure of the distortion or asymmetry of the distribution. A normal distribution is symmetric and has the skewness of 0, but in real-world scenarios, we are more likely to experience a non-zero skew. If the left tail of the distribution is longer, the distribution has a negative skew, whereas if the right tail is longer, the distribution as a positive skew.

For a feature without bias, we would expect to see a similar shape for either of the two classes, and hence the difference in the skewness of the two distributions would be 0 or close to it.

Kurtosis: Kurtosis is the measure of tail extremity, or the extreme values in either tail. A distribution with higher kurtosis will have longer tails as compared to a distribution with smaller kurtosis.

Hence, for a feature without bias, the difference in the kurtosis of the distributions for the two classes will be 0 or close to it.

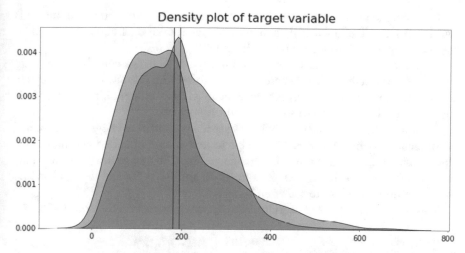

Fig. 3.3 Density plot for credit score of the two classes for feature "married"

For our credit score example, the differences between the three measures for the two distributions are as below:

Mean difference of target value between two groups: 14.4 (since the values are normalized, this translates to 14.4% difference in the credit score of two groups)

Skewness difference of target value between two groups. −0.7214

Kurtosis difference of target value between two groups: −0.9919

These metrices highlight a stark difference between the two distributions. The difference in the mean confirms that this is a sensitive feature and the underlying discrimination against the protected class in the dataset needs to be addressed. In order to better judge the difference and understand whether the distribution is really significant, it's important to look for skewness and kurtosis that would highlight presence of outliers. A large difference in skewness of the two distributions will indicate a big disparity in where the majority of two groups lie, and a large difference in kurtosis can indicate that one group has significantly more outliers than the other group – which would need to be addressed as part of the mitigation technique. This can either require outlier removal pre-processing or need the data to be examined further. In a few cases, it may be ideal to create smaller sub-groups for better analysis.

In an ideal scenario, the mean, skewness and kurtosis difference for the two distributions will be 0. For practical applications, the business analysts will have to advise the product owner on the threshold for the mean difference and then work to reduce the mean difference while reducing further the skewness and kurtosis difference.

When the Y Is Binary and S Is Continuous

Going back to our example of predicting the default on the loan, one of the protected features is the debtor's age. To handle a continuous feature like age, we can split it into bins and then do the analysis.

One approach is to create iterative bins to find the bin with the maximum SPD and DI. For Y = likelihood of default and S = Age for our dataset, the SPD and DI values for different bins are shown in Table 3.2:

For our example, we can see that creating a bin with an age cutoff does not help us. The issue with the approach here is the bin size – with just one bin, it is tough to identify the specific age groups that face the bias. Also, a larger bin size may hide the discrimination that happens for a specific age group.

To explore the data further, we will now see if any additional insights are revealed if we create multiple age bins. This helps us avoid large bin sizes and find out if discrimination is significant in specific age groups. The age bins we will test now are age < 20, age 20–30, age 30–40, age 40–50 and age > 50, and the SPD and DI values for different bins are shown in Table 3.3.

This reveals that we need to address the disparate impact for the age 20–30.

Additional Thoughts

When it comes to measuring discrimination, SPD and DI are always preferred route. But then in a few cases (or use cases), they may not be enough. In such scenario, these tests can be complimented with some statistical tests. For example, in our data on loan application, we want to check whether married customers who applied for a loan were discriminated against. To do this, we will consider the proportion of married applicants that were given a loan versus the proportion of non-married applicants that were given a loan (favourable outcome in both the cases). A statistically significant difference in these proportions would indicate discrimination.

The actual (Y) proportion of married that were hired can be represented as m and the proportion for the non-married as nm.

The statistical approach for testing whether non-married are discriminated would be if a one-sided null hypothesis $h0 : nm \geq m$ can be rejected. Simply put, if the hypothesis gets rejected, the probability is high that there is discrimination.

A simple test that can be further used here is a two-sample t test. One can argue to opt for a chi-square test of independence as well. The test statistic that would be used for a two-sample t test

$$\frac{\overline{m} - \overline{nm}}{\sqrt{\dfrac{s_1^2}{n_1} + \dfrac{s_2^2}{n_2}}}$$

Where s_1 and s_2 are standard deviation and n_1 and n_2 are frequency of m and nm groups, respectively.

The value obtained above thus can be used to calculate p-value for significance test.

Table 3.2 SPD and DI values for various age bins

Age bin cutoff	SPD	DI
20	−0.0382	1.0527
40	−0.0724	1.0991
50	−0.0616	1.0821
60	−0.0662	1.0873

Table 3.3 SPD and DI values for various age groups

Bin age group	SPD	DI
< 20	−0.0046	1.0060
20–30	−0.0961	1.1394
30–40	0.0060	1.0079
40–50	0.0494	1.0657
> 50	0.0596	1.0793

Conclusion

Key Takeaways for the Product Owner

The biases in the data are a result of conscious and subconscious human biases. Any process automated using this data or a prediction model built using this data is likely to have the same biases creep into the model. In this chapter we have covered how to identify the sensitive features that help identify and highlight the biases in place.

Identifying the sensitive features should be a priority and should be covered as early in the sprint planning as possible. This will be key in defining the stories that focus on removing the bias and in defining the problem statement that the data scientists will be working on.

An important decision at this stage is to determine the sensitive features, identify the advantaged/disadvantaged groups and define the thresholds for statistical parity difference and the disparate impact. You can choose to start with the same thresholds for all features and adjust them as needed – based on the business need, alignment with the strategy and the level of bias. We have used 0.1 as the threshold for SPD and 1.1 for DI.

Key Takeaways for the Business Analysts/SMEs

The techniques explained in this chapter are useful in identifying the sensitive features, but to understand their impact, the role of business analysts is critical. As seen in the chapter, there are a couple of features where the disparate impact exceeds the

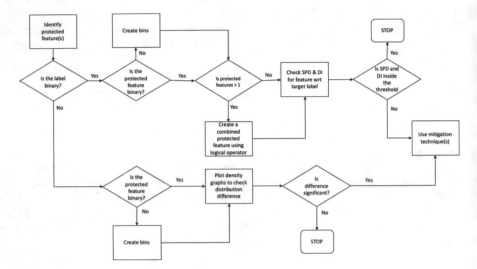

Fig. 3.4 Finalizing protected features

threshold, but we have not prioritized them for the analysis. This is because there are other features with much more significant impact, and this was the prioritization needed.

As a business analyst, it will be your role to work closely with the data scientists and advise the product owner on the prioritization needed. To summarize the journey you will be taking your product owner and the data scientists on, we have created the decision chart (Fig. 3.4) that can help you decide the techniques to use:

Key Takeaways for the Data Scientists

To the data scientists in the team, we recommend using the techniques laid out in this chapter and the associated code to help identify the sensitive features, the advantaged and disadvantaged groups and the data science stories that need to be completed for the feature to be built.

SPD and DI are just two metrics we have found that can be useful in most of the scenarios. However, as covered under the additional thoughts section, there are more tools available to you if you face more complex situations.

Bibliography

Dwork, C. et al. (2012) "Fairness through awareness," in Proceedings of the 3rd Innovations in Theoretical Computer Science Conference on - ITCS '12. New York, New York, USA: ACM Press.

Wikipedia Contributors (2019). Disparate impact. [online] Wikipedia. Available at: https://en.wikipedia.org/wiki/Disparate_impact [Accessed 18 Oct. 2019].

Chapter 4
Explainability

Introduction

The need to be able to explain is part of how we understand – from functioning of a machine to our actions. When faced with a decision-making process that does not explain its decisions, it is a natural human reaction to not trust it. This need for explanation to trust a decision is amplified when it is an algorithm making a decision for us. A large portion of the adoption for XAI has come from the regulated industries, where the need for explaining the decisions is not just good for client satisfaction and business but helps win the trust of the regulators too. As a result, eXplainable AI (XAI) has been around for a while, and more progress has been made in this area than other areas of creating a fair and ethical AI. In order to explain, XAI generally takes one of the three approaches: Creating local surrogate models that can explain an outcome of a black box model, feature explanation works towards explaining the relationship and impact of a given feature on the predicted outcome and then finally we have explainable models (or glass box models) where the model you are using for making the predictions is explainable and does not need the first two mechanisms to make its predictions interpretable.

The surrogate models are used to explain black box models because a black box model does not offer any explanation for its predictions. The approach usually attempts to explain the decisions by creating an interpretable model in the neighbourhood of the prediction. This is useful if you have to use a black box model or if you already have a model in production and need to explain the decisions. Another approach that can help explain decisions if you already have a model in the production is feature explanations. They work by trying to determine the sensitivity of the output to a feature at a time. The glass box models, as the name suggests, offer the other end of the spectrum as compared to the black box models. A model that is fully interpretable and ready for production deployment is a glass box model. In this chapter, we will look at the various techniques that belong to one of these approaches.

S. Agarwal, S. Mishra, *Responsible AI*,
https://doi.org/10.1007/978-3-030-76860-7_4

In addition to this, we will also approach explainability from a different point of view when we talk about how counterfactual fairness can be applied to "explain the model" by finding out what features need to have a different value and by how much for the prediction to change. This can be very useful when you want to expose the decision-making process to your users and want to provide them a feedback of what they should do for a decision to change.

We realise that in an ideal scenario, we would only have glass box models – models that are good to go live on a production environment and are also fully interpretable. However, given the recent advances, black box models can offer performance gains that are difficult to set aside. These models can have a wider reach and impact if product teams are able to explain better the predictions made by these models – regardless of the algorithm behind the black box model. The explainability that the business and the customers seek is not limited to a simple framework. Stakeholders are instead asking for a wider explainable canvas – and to achieve a much better understanding of how the input features impact the output. They feel more comfortable when the XAI matches the explainability that a linear model provides. Having said that, it may not be always possible to come out with explanation of a complex nonlinear model that matches the explainability of a simple linear model.

This is what we are going to try and challenge in this chapter. The standard approach to explaining the models is to create a model and then work on the explainability. We'd like to propose considering the explanation even before the model has been trained. We have already discussed the concepts of fairness and the bias in the data, and the next step is to move further into the data science lifecycle and look at what XAI can offer us. Considering XAI this early on will help us understand the features that would be used to eventually build the model.

We have grouped the techniques covered in this chapter into the four groups, based on the approaches covered earlier. Similar to other chapters, some of the techniques covered in the chapter are available in standard packages, but most of them are based on recent research or are taken from other areas that we feel can have an impact on the explainability. For the purposes of this chapter, we will represent the favourable outcomes with $Y = 1$ and unfavourable outcomes with $Y = 0$.

Feature Explanation

Information Value Plots

Information value plots is a technique that helps us understand how a feature impacts the output and is widely used for feature selection in logistic regression models. The underlying concepts of the IV Plots are weight of evidence (WoE) and the information value (IV). WoE helps us quantify the predictive power of a feature on the output, and the IV uses WoE to assign a score (IV) that can be used to compare and prioritize the features. Before we look at how to compute the IV, let's look at what the final value tells us about the feature's influence on the output and why it is important (Table 4.1).

This simple technique can help us compare the different features in a single scale and prioritize them based on their influence on the output. To calculate the WoE, we begin by splitting the feature into bins; the number of bins can vary, but we want each bin to have at least 5% of the values. We will represent the bin for feature X_j by $B_1...B_i...B_k$. Now the WoE for X_j in the bin i can be calculated as

$$WOE_{ij} = \log \frac{P\left(X_j \in B_i \mid Y = 1\right)}{P\left(X_j \in B_i \mid Y = 0\right)}$$

Where $P(X_j \in B_i \mid Y = 1)$ is the probability of a record having a favourable outcome ($Y = 1$) if X_j is in the bin B_i.

Using the WoE values, we can now determine the IV for the feature X_j (IV_j) over all the bins:

$$IV_j = \sum \left(P\left(X_j \in B_i \mid Y = 1\right) - P\left(X_j \in B_i \mid Y = 0\right) \right) \times WOE_{ij}$$

We can then iterate through all the features of interest and calculate the IV for them. The table earlier in this section can then be used to compare the IV for different features to understand the impact that they have on the outcome.

Figure 4.1 shows how the feature "interest rate" is related to the target feature – as the interest rate increases the probability of default increases.

However, as seen in Fig. 4.2, after building a bagging model, the trend between interest and the prediction probability of the bagging model differs from the trend shown by actual data. It can be seen that for lower values of interest variable, the prediction probability is at the lower end of default and then suddenly increases for a bin. This indicates that while the actual data indicates a possible relationship between the feature and the outcome, analysis of the predictions made by our model shows that the influence of this feature is weak on the output, with only one bin having an IV that crossed "strong influence" level. Therefore, it looks like it is not adding much value to the model, but then it will be too early to conclude anything as this only tells us that there is no strong linear relationship between the feature and

Table 4.1 Influence of the information value on the output

IV	Influence on the output
< 0.02	No influence or not useful for modelling
0.02–0.1	Weak influence
0.1–0.3	Medium influence
0.3–0.5	Strong influence
> 0.5	Suspiciously strong influence

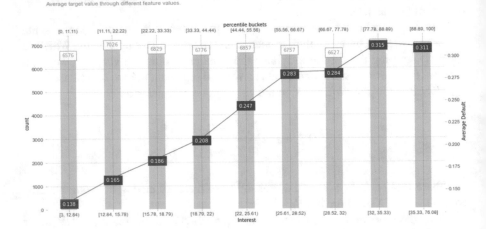

Fig. 4.1 As interest rate increases, the actual rate of default increases – showing the strong relationship between the feature and the output

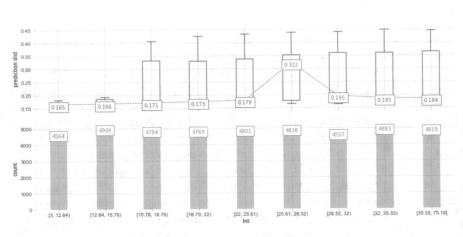

Fig. 4.2 As interest rate increases, the predicted probability of default does not change significantly except for one bin

the outcome, but we still need to investigate for any possible nonlinear relationship.

We understand that the information value plot may not always agree with the feature selection of a nonlinear bagging/boosting/stacking models and would sometimes be not enough to explain the trend as well. However, it can't be denied that IV plots are one of the first visual explanations that a stakeholder would appreciate. IV plots are typically linear in nature, and linear relationships are the easiest to explain.

Partial Dependency Plots

The next widely used and preferred method for feature explanation is the partial dependency plots (PDP). The PDP shows the relationship between the outcome and the feature being investigated. It creates a plot (Fig. 4.3) showing the relationship of the feature with the outcome across a range of values that are present in the data. For each value of the feature, the model returns the predictions for all the values of other features, which are then averaged. The benefit of this approach is that it is model-agnostic and can be implemented for any kind of classification or regression models. However, this approach assumes that the features are not correlated with each other. This is often not the case in real life, making it difficult to apply PDP. Another challenge in this approach is that it can use impractical values (Age, 22 years, and Work Experience, 20 years) that can lead to issues.

To compute the partial dependency, let X_s be the variable of choice and X_c be all the other variables in the data set. After fitting the model f, we get a prediction $\hat{f}$. Thus, partial dependency would be

$$\widehat{f_s} = \frac{1}{n}\sum \hat{f}(x_s, x_c)$$

Here we see that the prediction probabilities increase steeply beyond 19% of interest for the bagging (random forest) model we fit initially, showing a strong relationship between the feature and the outcome beyond this point. For the smaller values of interest rate, there seems to have no impact on the prediction probabilities. We

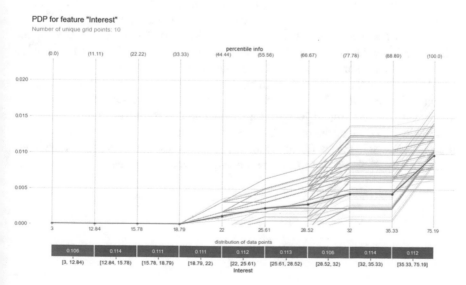

Fig. 4.3 Partial dependency plot for interest rate

can now see that interest rate has a relationship with the outcome, but the relationship is not a linear one.

Since in PDP all other features values are held constant, there are chances that sometimes PDP may not reflect reality. As it is assumed that the feature of interest is not correlated with other features, PDP is not very intuitive for a complex relationship. For the partial dependency of one of the features, e.g. interest rate, we assume that the feature "total liability" is not correlated, which is not a valid assumption. Similarly, for the computation of the PDP at a certain interest rate, we average over the marginal distribution of total liability, which might include values that are unrealistic for a particular value of interest. Also, as PDP is computed based on average predictions, if the variable in investigation has both positive and negative impact, the plot would be almost a horizontal line (cancelling -ve and + ve values), which brings us to the next topic on accumulated local efforts that aims to overcome these shortcomings.

Accumulated Local Effects

As discussed above, PDP has challenges when the features are correlated, as during extrapolation it can create instances that are not practical and also outside the distribution of training data. On the contrary, accumulated local effects (ALE) plots can provide with sensible insights even if the predictors are correlated. Since ALE partially isolates the effects of other features, to a large extent, it is not impacted by other correlated features. ALE uses the conditional distribution of the feature and creates more realistic data and compares the data with instances from the dataset that are similar in nature and it calculates the average of model prediction differences compared to PDP's method average over predictions. For instance, to calculate effect of a feature at a particular value, ALE considers the average of the difference in model predictions of the values in the vicinity of that particular value (all observation around that one value in consideration).

ALE plots calculate differences in predictions instead of averages. For instance, for interest rate at 70 (70 percent), it would create a local area and find out all customers with interest around 70 and let the model predict for interest rate between say 69 to 71 and take the difference of the predictions. The graph (Fig. 4.4) shows us that there is a change in the predicted outcome which increases consistently for the interest rate above 35% and that the steepest change happens between 27% and 30%.

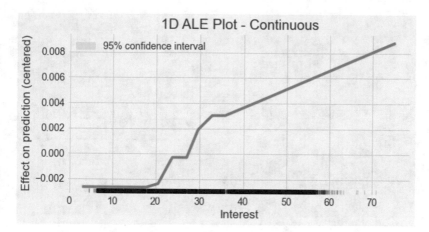

Fig. 4.4 As the interest rate increases, the difference in predictions in the neighbourhood of the interest rate also increases

Sensitivity Analysis

The techniques for feature explanation covered so far are available as standard packages. We would now like to propose a new technique that can be used for feature explanations – sensitivity analysis. This technique covers all values between the minimum and maximum values for the feature in the dataset, demonstrates the sensitivity of the model at the boundary conditions, average and for other values in between and finally the resulting plots are easily interpretable without the need to understand the underlying math – making it a strong tool to use when explaining to the user or the business stakeholders.

In sensitivity analysis, we use the black box model to predict the outcome as we hold all other features at constant either at their mean, min or max (or may be at interquartile range) and loop through all values of the feature under investigation from its minimum to maximum. Hence, we are able to have an insight that talks about how a model would predict a value of one particular feature when all other features are held constant. This helps us understand the value(s) of that particular feature when the model would be very sensitive and would show a dip or a peak, revealing sensitivity values of a particular feature. A sudden increase or decrease in prediction value with a tiny change in data may question the model and may require the value to be treated in some other manner.

$$\text{Sensitivity}^k_{mean} = f\left(x^k_s, \overline{x_{c1}}, \ldots, \overline{x_{cN}}\right) \; for \; k = x^{min}_s, \ldots, x^{max}_s$$

where x_s is feature under investigation
x_c are all other features

In Fig. 4.5, the plot for the feature interest rate, we see that the prediction probability starts decreasing slightly after 20% in steps until 40% when other features are held constant at their mean. The trend seems quite uniform, and there seems to be no sudden

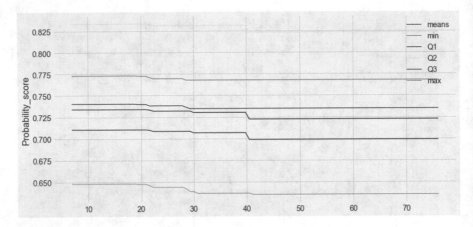

Fig. 4.5 Sensitivity analysis of interest rate keeping all other features fixed

increase or decrease in prediction probability for any particular value(s) of interest variable. When other features are held constant at a different value (minimum, maximum, Q1, Q2 and Q3), the sensitivity of the interest rate does not change significantly.

The python implementation of this method is as follows.

```python
def get_data(d):
    return d[:, 0]
```

```python
analyser = ans.Analyser(clf, data_stats1, clf.predict_proba, get_data)
```

```python
res = list(analyser.run_on_features(list(col_val), 100, multicore=False))
```

```python
%matplotlib inline
plt.style.use('seaborn-white')

plt.style.use('fivethirtyeight') # Good looking plots
pd.set_option('display.max_columns', None) # Display any number of columns
from matplotlib.pyplot import *

for feature, result in res:
    print(feature)
    points = [r[0] for r in result]
    means = [r[1][0] for r in result]
    mins = [r[1][1] for r in result]
    q1st = [r[1][2] for r in result]
    q2nd = [r[1][3] for r in result]
    q3rd = [r[1][4] for r in result]
    maxs = [r[1][5] for r in result]

    plt.figure(figsize = (12,6))

    plt.plot(points, mins, linewidth=1, color='red')
    plt.plot(points, means, linewidth=1, color='green')
    plt.plot(points, q1st, linewidth=1, color='blue')
    plt.plot(points, q2nd, linewidth=1, color='yellow')
    plt.plot(points, q3rd, linewidth=1, color='black')
    plt.plot(points, maxs, linewidth=1, color='olive')
    plt.ylabel("Probability_score")

    plt.legend(["means", "min", 'Q1', 'Q2', 'Q3', 'max'], loc=1)

    plt.show()
```

The above code calls the following function:

```
from collections import OrderedDict
from multiprocessing import cpu_count, Pool
import numpy as np

_N_CORES = cpu_count() - 1

class _SAResult:
    """ Storage for Sensitivity Analysis results"""
    def __init__(self):
        self._ts = OrderedDict()

    def add(self, value, df):
        """ Add returned probabilities for a given value

        Args:
            value (:obj:`number`)
                Value of the feature the predictions were made on
            df (:obj:`iterable`)
                Probabilities
        """
        self._ts[value] = df

    def __iter__(self):
        return (v for v in self._ts.items())
```

```
class Analyser:
    """ Perform sensitivity analysis on any model in sklearn

    Args:
        model (:obj:`sklearn.Model`)
            The model object
        df (:obj:`pandas.DataFrame`)
            Result of running DataFrame.describe() on the target dataset.
            The 'std' row should be dropped before running.
        predict (:obj:`function|method`)
            Function for running the prediction on. Example: model.predict
        get_result (:obj:`function|method`)
            Function for filtering the raw results as returned by the model
            upon running predict.
    """
    def __init__(self, model, df, predict, get_result):
        self._model = model
        self._df = df
        self._predict = predict
        self._get_result = get_result

    def run_on_features(self, features, steps=100, multicore=True):
        """ Run sensitivity analysis on a set of features.

        Args:
            features (:obj:`iterable`)
                Features to analyse. All values should be strings.
            steps (:obj:`int`, :optional)
                Number of data points between min and max to predict on. Default is 100.
            multicore (:obj:`bool`, :optional)
                Set this to False to run the predictions on a single process.

        Returns:
            Iterable of tuples (feature, results), where results is a _SAResult object
        """
        if multicore:
            args = ((feature, steps) for feature in features)

            with Pool(_N_CORES) as pool:
                results = pool.starmap(self.run_on_feature, args)
        else:
            results = []
            for feature in features:
                results.append(self.run_on_feature(feature, steps))

        return zip(features, results)
```

```
    def _get_step_values(self, feature, steps):
        fd = self._df[feature]

        return np.linspace(min(fd), max(fd), num=steps)
```

Model Explanation

Split and Compare Quantiles

Before going through the detailed model explanation techniques, we want to start with a very simple but effective way to understand the output of the model. This analysis helps explain to the business stakeholders the errors made by the model and hence is a powerful tool in the hands of a product owner to make decisions about the model. The split and compare quantiles helps us define a decision threshold for a classification problem by giving a clear understanding of the impact of our decision on the confusion matrix and evaluate if the model helps meet the business objectives.

All models, however good, come with some error in their predictions. Therefore, even the most conservative outlook will still have some errors in the prediction. In our example when we try to identify the debtors who are not likely to default, we will have a small number of defaulters – the error in the prediction. Similarly, as we try to identify the most likely defaulters, we will also incorrectly identify some debtors who won't be defaulting on the loan.

To implement it, we divide our dataset into equal quantiles and then split it by favourable and unfavourable outcomes. In the code below, we have split the sample data first into deciles and then by the label. The block of code below cuts the data into ten equal size bins and then calculate the percentage of observations (over the entire data) in each bin for both the labels.

```
def ret_tag(val):
    for i in range(10):
        if val in categories[i]:
            return tags[i]

def zero_list_maker(n):
    list_of_zeros = [0] * n
    return list_of_zeros
```

```
cuts_by_quantile = pd.qcut(view['Score2'], 11, duplicates='drop')
toplt = cuts_by_quantile.value_counts().index.categories.mid
categories = cuts_by_quantile.value_counts().index.categories
left_bounds = categories.left
right_bounds = categories.right
tags = range(1, 11)
```

```
def bar_width_per_quantile2(view, label_filter):
    view_filtered = view[view['Default'] == label_filter]
    view_filtered['tag'] = view_filtered['Score2'].map(lambda x: ret_tag(x))
    len_view_filtered = len(view_filtered)
    view_filtered['ID']=view_filtered.index
    count_pertag_filtered = view_filtered[['ID', 'tag']].groupby(['tag']).count().rename(columns={'ID': 'Count'})
    count_pertag_filtered['perc'] = count_pertag_filtered['Count'] / len(view) * 100
    sc = count_pertag_filtered['perc'].tolist()
    sc = zero_list_maker(10 - len(sc)) + sc #10
    return sc
```

The code below splits the decile data by the labels and prepare the data for creating the plot.

```
scl = bar_width_per_quantile2(view, 1)
sc0 = bar_width_per_quantile2(view, 0)
```

```
name = [str(left_bounds[i])+'-'+str(right_bounds[i]) for i in range(10)] #10
```

```
data_p = [(name[i], scl[i], sc0[i]) for i in range(10)] #10
```

```
df = pd.DataFrame(columns=["Range","Label: 1", "Label 0"],
                  data=data_p)
```

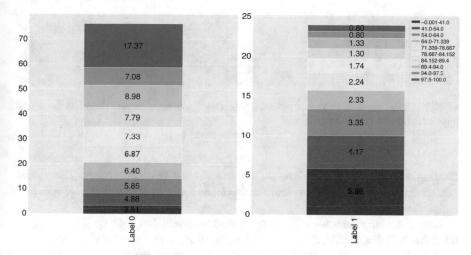

Fig. 4.6 The actual favourable and unfavourable outcomes split into deciles by the predicted probabilities

The chart (Fig. 4.6) shows the favourable and unfavourable outcomes split into deciles by the predicted probability. Let's assume that the predicted probability higher than 97.5% is identified as not likely to default. This means 18.17% of the debtors would be identified as not likely to default. Of this 0.8% are the ones that will default, whereas almost 60% of the total debtors that are not going to default will be identified as likely to default. The first part represents potential commercial loss by incorrect labelling of defaulters as not likely (false positives) and the second part represents potential loss of opportunity (false negatives).

This should be used by the product owner and the business analysts to determine the optimum decision boundary that helps minimize the false positives and the negatives.

Global Explanation

For a black box model, the feature explanations and understanding the output's impact on the users is winning half the battle. But it still doesn't mean that we've been able to peek under the hood and take a look at what is really happening inside.

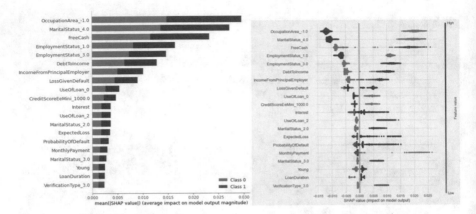

Fig. 4.7 Global explanation output by SHAP showing the average impact (left) and the actual impact (right) of a feature on the model

Starting with global explanation, we want to discuss two approaches. The global explanation tries to linearly explain the model as a whole, whereas the local explanation does it in parts.

Global explanation is widely practiced, and libraries like LIME or SHAP support it too (we will discuss SHAP given its wider coverage and the better visual insights). SHAP allows explanation for a tree-based model, linear models and Kernel-based method. As our base model is random forest, we would stick to tree-based SHAP explainer. Tree models are conceptually different than an additive linear model. The Shap TreeExplainer thus estimates the Shapley value using tree nodes.

The plot (Fig. 4.7) on the right shows the overall strength of features along with class separation. The features higher on the chart have clear impact on both the labels. The feature importance chart (Fig. 4.7) on the left is generated after traversing across all the instances and shows the average Shap value for each feature. This provides a much simpler view to understand the relative importance of the features on the output labels – features with higher rank have an average higher impact.

The next few plots (Fig. 4.8) are typical dependency plots, as introduced in the first part of this chapter. Here, the dependency plot is in the form of a scatter plot reflecting relationship between one single feature (the feature being investigated here is interest) and the response feature. Here x axis represents value of the feature interest, while the y axis is the SHAP value for that feature. The plot on the right is a similar plot but additionally represents interaction with another feature.

The below three plots (Fig. 4.9) show interaction of the feature "interest" with three other most interactive feature. Thus, we can not only infer independent impact of a feature on the response variable but also look at collective impact through interactions. Moreover, this plot can further highlight the kind of interaction (inverse or direct or nonlinear) and the strength of interaction.

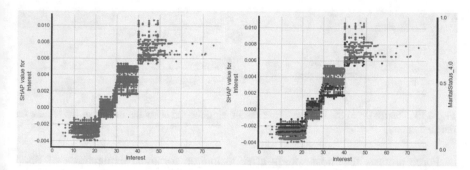

Fig. 4.8 Scatter plot showing the relationship between interest and its Shap value (left) and along with another feature Marital Status = Divorced (right)

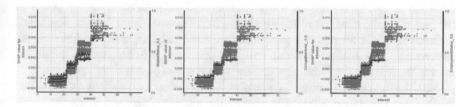

Fig. 4.9 Scatter plot showing the relationship between interest and its Shap value along with one other feature at a time

Local Explanation

Feature importance is unarguably the most important metric that is assessed during/after model development. Perhaps even more so than the overall model performance as it relays the rationale behind the outcome of the model, and without that, understanding the performance itself may not have much meaning or the support to get to a production environment. The feature importance is even more pertinent for a non-linear or a black box model, as for these models, it is very tough to derive coefficient of features compared to a linear model. A lot of times a squashing function (sigmoid or logistic or arctangent or hyperbolic tangent or Gudermannian function) is used to bring in the data from a higher dimension to lower dimension and squash the values between a bounded region and that makes it increasingly difficult to understand what the model is doing. In a typical classification algorithm, a bounded space is between 0 and 1.

The global explanation is a generalist explanation; it gives us an averaged view over the full range of the data but does not tell us what is happening locally at any point. Sometimes it is important to deep dive into local explanation to be able to better understand the model's behaviour, especially for edge cases and cases that requires special attention.

Fig. 4.10 The length of the individual bars and their direction show the impact of the feature on the prediction

Force plots is the most sought-after technique for local explanation of Shap library. It shows both the direction and magnitude of the feature and how they drive the outcome. Take for instance Shap's force plot shown in Fig. 4.10. In the illustration (Fig. 4.10), we investigate a customer whose probability of non-default is 0.8256173 (and default is 0.1743827). The colour of the individual bar shows the direction of the feature impact while the length of the individual bar the power of the impact. In the case below, the maximum impact comes from "occupation area". Remember this is for this one particular customer and should not be generalized. Interestingly, the distance between base and prediction will be equal to the difference in the length of the blue bars and the length of the pink bars.

Just like the above-discussed force plot, SHAP offers an alternative view called the decision plot. Given interaction with multiple stakeholders and customers, decision plots seem to be more favourite for a layman. The very reason that one needs a decision plot over a force plot is the visualization capability. A decision plot easily shows a large number of feature and their impact level. If you compute the decision plot with multiple instances, it would be interesting to note the feature-wise outliers that would pop out in the plot.

The x axis denotes the model's output which in this case is the probabilities of default. The y axis lists the model's features with their actual values represented in parenthesis next to the plot. The plot further (Fig. 4.11) gives insight into how each feature contributes to the final outcome more like an additive model where adding each feature value leads to final outcome. You can further group similar prediction paths using hierarchical clustering method in the same plot.

The next interesting plot (Fig. 4.11) is the waterfall plot that visually tell you the story of each feature. It vividly speaks about the negative and positive impact (colour of the bar), their strength (length of the bars) and how they shake the prediction from the previous feature. Features here are arranged in descending order, and all less important features are clubbed together as "other features" at the bottom. This is again used for local explanation and should not be generalized.

Figure 4.12 shows an example that plots the explanation for a particular customer. The values on the x axis are probabilities of default. The grey text before the feature names shows the value of each feature for this sample. It is interesting that how this matches the force-plot as shown in Fig. 4.10. The two plots (Fig. 4.12) are

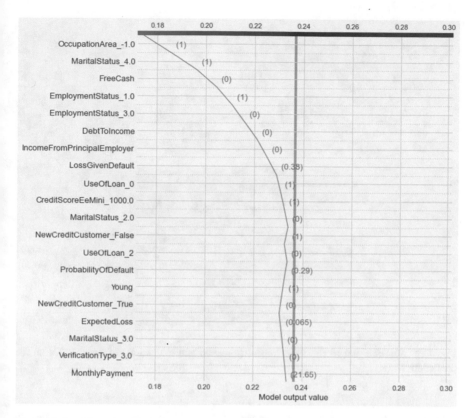

Fig. 4.11 Decision plot showing how each feature's contribution adds up leading to the prediction

one and the same thing as one is inverse of other as the first plot shows default probability and the next one is for non-default.

The bars in the plots show the impact; the lengthier the bar, the more the impact. The colour of the bar depicts the direction in which that particular feature impacts the final probability.

The next method (Fig. 4.13) is about interpreting a package from Microsoft which is based on generalized additive models with pairwise interactions (GAM is discussed in more detail later in this chapter). The package implements explainable boosting machine and is quite simple to use and useful during model development and intermittent explanations. It has a wide coverage with global explanation, local explanation and sensitivity all included. Unlike other methods, GAM allows to track non-linear pairwise interaction between features as well.

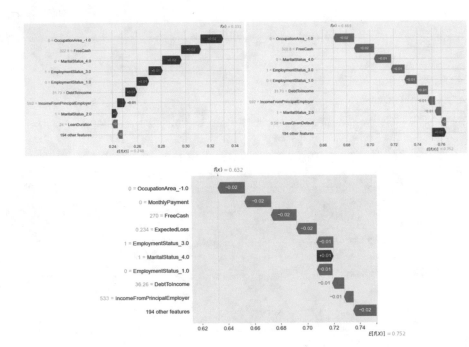

Fig. 4.12 Waterfall plot to show the impact (direction and magnitude) of the features on the prediction

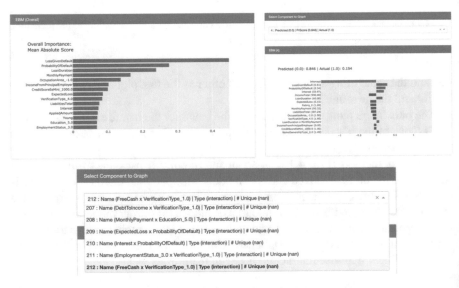

Fig. 4.13 (1) EBM overall average feature importance, (2) local explanation for observation no. 4, (3) feature importance score choice for interaction features

Morris Sensitivity

Morris sensitivity is a global sensitivity analysis technique that is equally intuitive for data scientists and the business stakeholders. It describes the impact that a feature has and how it would change the prediction, giving a high-level view if a feature has any effect on the model output – helping the analysts isolate the feature and investigate further. In some cases, it can even help the analysts discard the feature if it helps result in less noisy model and reduce the feature space.

The sensitivity analysis (Fig. 4.14) determines the feature importance by changing the feature values (or ignoring them) while all the other features are constant and computing the predictions using them. Morris sensitivity works by changing the value of the feature under consideration in one direction (δx_i) and looking at the difference in prediction caused by that change. This action is performed iteratively for the full range of the feature under consideration and then aggregated over all the differences. Since the negative and the positive values will cancel out, we need to take the absolute value of the square difference of the value.

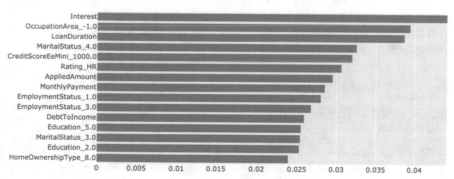

Fig. 4.14 Morris sensitivity plot showing the feature sensitivity

Explainable Models

Generalized Additive Models (GAM)

Linear models are inherently explainable but, as the name implies, do not capture nonlinearity. Generalized additive models are a nonlinear version of the generalized linear models and provide an explainable and flexible approach like linear models while capturing nonlinearity. GAMs don't restrict the function to be merely additive but allow additive model with functions for each feature. Mathematically, the relationship in a GAM looks like this:

$$g(E[y \mid X]) = \beta_0 + f_1(X_1) + \ldots + f_M(X_N)$$

For the link function for logistic regression

$$\log\left(\frac{P(y=1 \mid X)}{P(y=0 \mid X)}\right) = \beta_o + f_1(X_1) + \ldots + f_M(X_N)$$

Since this is an additive model with a separate function for each feature, the impact of each feature is known at all times. There are no hidden inner workings of the model making the explanation difficult. This ability makes this a "glass box" model and deserves a strong consideration if your other options so far are black box models with additional explanation added.

GAM uses splines for learning the link function and also computes the weights for the splines. Additionally, it can also use a regularizer to reduce overfitting and the grid search method to optimize the weights. In the illustration (Fig. 4.15), we can see the accuracy of the model to be quite high (without any overfitting). Interestingly, functions for predictors can also be automatically derived during model training and you don't need to state the function before training the model. Just like other linear models, one has an option of either declaring the smoothing parameters or use cross validation to find the optimal smoothing parameters.

```
gam.accuracy(X_train.loc[:,p], y_train)
```

```
0.7809384027583636
```

```
gam.accuracy(X_test.loc[:,p], y_test)
```

```
0.7802359080284829
```

```
gam.summary()
```

```
LogisticGAM
=============================================    =================================================
Distribution:             BinomialDist Effective DoF:                                     83.2995
Link Function:              LogitLink Log Likelihood:                                 -20180.3891
Number of Samples:              42924 AIC:                                            40527.3772
                                      AICc:                                            40527.713
                                      UBRE:                                               2.9457
                                      Scale:                                                 1.0
                                      Pseudo R-Squared:                                   0.1411
=============================================    =================================================
Feature Function          Lambda               Rank         EDoF         P > x        Sig. Code
=============================================    =================================================
s(0)                      [0.6]                20           2.0          0.00e+00     ***
s(1)                      [0.6]                20           1.0          9.62e-01
s(2)                      [0.6]                20           2.2          4.33e-09     ***
s(3)                      [0.6]                20           1.0          6.62e-07     ***
s(4)                      [0.6]                20           1.0          9.59e-01
```

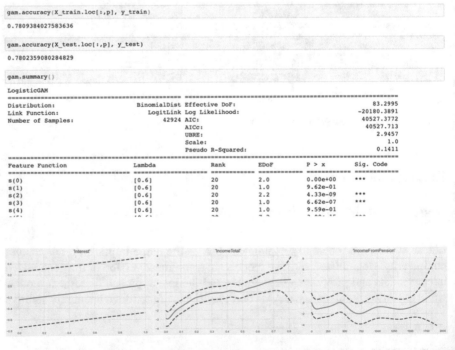

Fig. 4.15 Three charts above show features ranging from linear (Interest) to highly nonlinear (income from pension) and how they have been handled separately and accurately by the GAM model

The summary can be read as any regression summary with coefficient as the power of the feature, p-value as significance and so on.

If you already have a black box model in the production, GAM can be used as a surrogate model to provide the explanation. We have seen multiple methods of peeking under the hood for a black box model, but none of those provides the flexibility of designing a custom locally explained model – which can provide an intuitive explanation for your stakeholders.

The example below shows (Fig. 4.16) how GAM can be used as a surrogate model(s) to explain the behaviour of a random forest model. Figure 4.16 shows the nonlinear boundary as identified by the random forest model. We are trying to explain the prediction for the instance under investigation.

The process of creating a surrogate model typically involves defining a neighbourhood around the instance under investigation using multiple distance algorithms and then training a local model for using all the instances in the neighbourhood as the training data. Since GAM is capable of handling nonlinear relationship, you can experiment with the size of the neighbourhood and then train the corresponding GAM models.

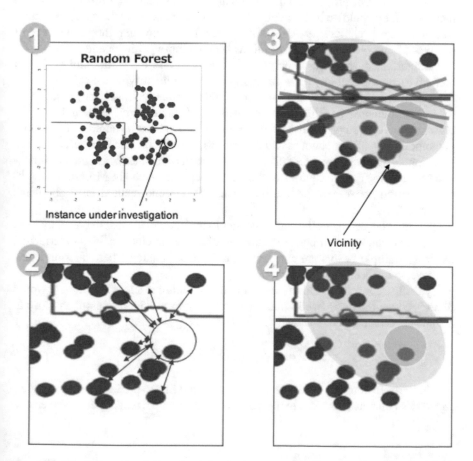

Fig. 4.16 Identifying the neighbourhood for the instance under investigation before training a local model to explain the predictions

The process usually involves training multiple models and then choosing the one that best approximates the performance of the black box model being explained. If you have identified multiple neighbourhoods and created multiple models, you can use the same model for all the instances within the neighbourhood. In case of overlap between the neighbourhoods, you can choose the model that has higher accuracy for the overlap to provide the explanation.

Counterfactual Explanation

We cannot finish the chapter on XAI without discussing the use of counterfactual explanation in explaining the model's behaviour. We have seen customers write back many a times to our clients where they've asked questions like: "Why me?", "What's next?" or "What if?". To answer questions like these, what the business really needs to understand is what would make the decision change – or what could have been different in the customer's data that could've meant a different outcome for them. A lack of proper explanation creates mistrust, but an effort to explain to them how they could've had a better outcome can create a much longer and more fruitful business relationship. Also, with the regulations giving rights to the customers to ask for explanation, these questions become more pertinent and valid.

Counterfactual explanation helps us achieve this. As the name implies, it is a method for countering the fact or reversing the fact or identifying the changes in the features that would change the outcome. Mathematically what change in X will inverse the outcome $\hat{Y}$. This allows the service provider to give explanation of the outcome and also suggest what needs to be done to get the favourable and desired outcome. Most of the counterfactual methods, CEML or Microsoft's DiCE, are quite model agnostic. We won't dive deep into each of the model as both are fundamentally solving the same purpose albeit using separate methods. Most of the methods discussed above explains the outcome, but this one method talks about the "What if?".

This method also helps the end-customer to understand to what degree they need to bring in change in their portfolio and don't leave them clueless. In counterfactual fairness, multiple factors are considered in decision-making where altering a few can reverse the decision.

Typically, a machine learning model aims to minimize a loss function. It would like to reduce the error (while training) by trying to predict an output as close to actual. This can be mathematically represented as

$$argmin_x \, l\big(f(x), y\big) + C$$

Similarly, in counterfactual, the algorithm would like to predict $\hat{Y}$ for a counterfactual x' so that the $\hat{Y}$ is close to the desired y'. The regularizer is a penalty to ensure the x and x' are close to each other, to ensure least deviation between x and x' and to provide a solution that implies least changes in x.

$$argmin_{x'} \, l\left(\hat{f}(x'), y'\right) + C.\theta \, (x', x)$$

where l is loss function

$\hat{f}$ is prediction function

x is counterfactual input

x' is counterfactual result

y' is requested counterfactual response

$C. \, \theta$ is some regularizer ($l1$ or MAD) to penalize the difference between x and x'

It would be a bit different for a tree-based or an ensemble model like bagging or boosting. In a tree-based non-ensemble model, counterfactuals can be found by listing all possible paths that lead to the desired outcome y', finding all leaves that lead to y' and then finding the minimal change in x that can lead to that leaf. Another approach is for the additional regularizer to include a penalty for sparsity in the difference between the x' and x and also a penalty to ensure that x' is inside training data manifold to ensure x' is realistic and is very near the training data and holds to training data correlations among the features.

Using ceml.sklearn in python, we use the same dataset as we were following in the chapter although with a smaller subset of features. After training a random forest model with 20 features, we tried to find a counterfactual decision for one instance.

```
from ceml.sklearn import generate_counterfactual
```

```
x = X_test.loc[1,p]
print("Prediction on x: {0}".format(clf2.predict([x])))
```

```
X_train.loc[:,p].columns
```

```
Index(['OccupationArea_-1.0', 'MaritalStatus_4.0', 'FreeCash',
       'EmploymentStatus_1.0', 'EmploymentStatus_3.0', 'DebtToIncome',
       'IncomeFromPrincipalEmployer', 'LossGivenDefault', 'MonthlyPayment',
       'ExpectedLoss', 'UseOfLoan_2', 'UseOfLoan_0', 'ProbabilityOfDefault',
       'VerificationType_3.0', 'CreditScoreEeMini_1000.0', 'Rating_HR',
       'MaritalStatus_2.0', 'MaritalStatus_3.0', 'Interest', 'WrExLess10'],
      dtype='object')
```

Let's consider one record in our dataset that has had an unfavourable outcome. If we are investigating this and want to tell the user what would have changed the outcome to a favourable one, we would want to apply the counterfactual explanation to make the determination. For this sample record, the original value of the features is displayed below. Since this algorithm requires the trained model as an input, the sample record we are using for this example was not used as part of the training or test data for the model.

```
x.values
```

```
array([1.00000000e+00,  1.00000000e+00,  0.00000000e+00,  0.00000000e+00,
       0.00000000e+00,  0.00000000e+00,  0.00000000e+00,  4.29111556e-01,
       6.34900000e+01,  5.92000000e-02,  0.00000000e+00,  1.00000000e+00,
       1.71474308e-01,  0.00000000e+00,  1.00000000e+00,  0.00000000e+00,
       0.00000000e+00,  0.00000000e+00,  1.75300000e+01,  1.00000000e+00])
```

The original prediction for this record was

```
x = X_test.loc[1,p]
print("Prediction on x: {0}".format(clf2.predict([x])))

Prediction on x: [0.]
```

The original prediction $\hat{Y}$ of X was a binary outcome value of 0 (corresponding to unfavourable outcome). The counterfactual function is highly customizable. A user can also specify the features that needs to be changed and not all features. For example, age of customer or marital status or no. of loans can be removed from feature wish list in order to ensure that the algorithms suggest no change in those features (so that the recommendation does not include inappropriate suggestions – like changing the marital status).

The counterfactual function gives us three outputs: The values for the features that would lead to the alternative outcome for the record (what needs to change), the new output as predicted by the model to confirm that the prediction would indeed be different and finally the difference between the original values for the features and the recommended values (how big is the change required).

```
print(generate_counterfactual(clf2, x.values, y_target=1, features_whitelist=None))

{'x_cf': array([ 1.00000141e+00,  9.99985689e-01,  8.64387829e-06,  5.00032865e-01,
        5.23191195e-05, -1.09242105e-05,  3.47642619e-08,  3.87700951e-01,
        6.34899003e+01,  1.07455501e-02, -1.82750739e-05,  9.99987645e-01,
        1.70945116e-01,  2.07879858e-05,  1.00000060e+00, -2.50219763e-06,
        1.17452767e-05,  8.59227794e-05,  1.75301229e+01,  1.00000010e+00]), 'y_cf': 1.0, 'delta': array([-1.40684237
e-06,  1.43108023e-05, -8.64387829e-06, -5.00032865e-01,
       -5.23191195e-05,  1.09242105e-05, -3.47642619e-08,  4.14106053e-02,
        9.96768158e-05,  4.84544499e-02,  1.82750739e-05,  1.23545218e-05,
        5.29192340e-04, -2.07879858e-05, -6.04826783e-07,  2.50219763e-06,
       -1.17452767e-05, -8.59227794e-05, -1.22943169e-04, -9.73650658e-08])}
```

```
# Feature 1
original_value = 1.00000000e+00
CF_value = 1.00000141e+00
print ('Delta: ', original_value-CF_value)

Delta:  -1.4100000000905055e-06
```

```
# Feature 2
original_value = 1.00000000e+00
CF_value = 9.99985689e-01
print ('Delta: ', original_value-CF_value)

Delta:  1.4311000000044594e-05
```

```
# Feature 8
original_value = 4.29111556e-01
CF_value = 3.87700951e-01
print ('Delta: ', original_value-CF_value)

Delta:  0.04141060499999999
```

A simple manual check shows that the deviation required in few features like occupation area and marital status is not required, but there needs a reduction in feature like loss given default value. Information like this can be incredibly useful to a customer in working to get a better outcome.

Conclusion

For any AI-driven product, the ability to explain its decisions, especially the ones made using intelligent models, is critical to gaining user trust and driving adoption. Being able to bring in the capability to explain, therefore, is a must have for any product owner today. However, at the same time, this ability cannot be added at the cost of performance of the intelligent models being trained or at the risk of cost overruns that make the product any less viable commercially.

Your choice of the technique will depend on multiple factors and is not limited to the ones covered in this chapter. An approach that has helped us every time in deciding on the right techniques is to use a questionnaire like the one below.

1. Why do you need explainability?
2. What is the coverage of your explainability?

    (a) Do stakeholders understand the algorithms and the manual process?
    (b) Do they know what is happening under the hood?
    (c) Do they understand the trade-off?

3. Who is the audience?

    (a) What do they perceive about the process and algorithm?
    (b) Will they lose trust if they don't understand the algorithm?
    (c) Are they the decision-makers?
    (d) Were any of their member involved in the design process?
    (e) What kind of explainability they are seeking?

4. Was the explanation enough and comprehensive?

    (a) Did you explain the data?
    (b) Did you explain the objective function?
    (c) What will be the BAU process?

In this chapter, we have tried to cover a wide range of techniques that can help you bring in explainability to your products using the mechanism that works best for your development team and your users and at the right cost. The chart (Fig. 4.17) shows how these techniques compare with each other in terms of the implementation complexity and the method explainability.

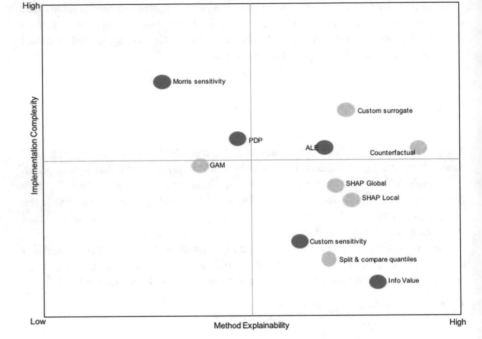

Fig. 4.17 Implementation complexity vs method explainability for the techniques covered in the chapter

Bibliography

Apley, D.W. and Zhu, J. (2020). Visualizing the effects of predictor variables in black box supervised learning models. Journal of the Royal Statistical Society: Series B (Statistical Methodology), 82(4), pp.1059–1086.

Artelt, A. and Hammer, B. (2019) "On the computation of counterfactual explanations -- A survey," arXiv [cs.LG]. Available at: http://arxiv.org/abs/1911.07749.

Lares, B. (2018). Machine Learning Results in R: one plot to rule them all! I R-bloggers. [online] Available at: https://www.r-bloggers.com/2018/07/machine-learning-results-in-r-one-plot-to-rule-them-all/ [Accessed 16 Apr. 2021].

Li, J. (2017). Introducing PDPbox. [online] Medium. Available at: https://towardsdatascience.com/introducing-pdpbox-2aa820afd312 [Accessed 16 Apr. 2021].

Li, Z. (2020) "Global sensitivity analysis of the static performance of concrete gravity dam from the viewpoint of structural health monitoring," Archives of

Computational Methods in Engineering. State of the Art Reviews. doi: 10.1007/s11831-020-09434-0.

Molnar, C. (n.d.). 5.1 Partial Dependence Plot (PDP) I Interpretable Machine Learning. [online] christophm.github.io. Available at: https://christophm.github.io/interpretable-ml-book/pdp.html.

Servén, D. (2020). pyGAM Documentation. [online] . Available at: https://readthedocs.org/projects/pygam/downloads/pdf/latest/ [Accessed 16 Apr. 2021].

Verma, S., Dickerson, J. and Hines, K. (2020) "Counterfactual explanations for machine learning: A review," arXiv [cs.LG]. Available at: http://arxiv.org/abs/2010.10596.

Wachter, S., Mittelstadt, B. and Russell, C. (2017). Counterfactual Explanations Without Opening the Black Box: Automated Decisions and the GDPR. SSRN Electronic Journal. [online] Available at: https://papers.ssrn.com/sol3/papers.cfm?abstract_id=3063289 [Accessed 14 May 2019].

Chapter 5
Remove Bias from ML Model

Introduction

In this chapter, we will be covering the techniques to overcome the bias in the data. We have picked up the techniques that are model agnostic and can be used with most of the machine learning algorithms. These techniques offer a lot of scope for optimization and the business analysts in the team will have to take the lead in understanding and deciding the best approach for the optimization that can be achieved.

In addition to the flexibility in choice of algorithm and the optimization, these techniques also help you explain the decisions using the features in the dataset and help make the final model much more explainable. There are off-the-shelf packages available; IBM's AIF360 and Microsoft's Fairlearn are just the two to mention in addition to others that can be used to assess and improve fairness, but we won't be covering them in this chapter. Instead of covering a package, we want to cover the techniques and show you how you can utilize them to mitigate the bias while not sacrificing the quality of the predictions. By the end of this chapter, you will be able to understand these techniques in more detail and how to choose between them for a given problem, what machine learning algorithms to use (e.g. classification or regression) and how to define the accuracy metric for your problem.

We will be looking in detail in this chapter the reweighting of data and additive counterfactual fairness. They are both very powerful techniques in helping overcome the data bias and complement each other in their ability to handle a single protected feature or multiple protected features and the support for algorithms. In the lifecycle of building a machine learning model, you can perform reweighting before any model training happens, whereas the ACF is applied during the model training. Based on the business problem at hand, by the time you are done with this chapter, you should be able to decide which approach meets your requirements and will work with the algorithm you choose for training the model. Let's look at them in more detail now.

© The Author(s), under exclusive license to Springer Nature Switzerland AG 2021
S. Agarwal, S. Mishra, *Responsible AI*,
https://doi.org/10.1007/978-3-030-76860-7_5

Reweighting the Data

As the name suggests, reweighting assigns weights to the data based on a protected feature. These weights are then used along with the input data for loss function optimization. A big benefit of this approach is that the data is not altered to achieve any reduction in the bias. Before we go into the details, the key features of the reweighting technique are the following:

1. It can handle one protected feature at a time.
2. Only for classification-based algorithm.
3. You can create composite features to handle multiple features together – composite features can be an incredibly powerful way to handle multiple features together, but the features you can combine will depend on the features and the problem at hand.
4. And finally, no dip in accuracy (or hardly any!).

The purpose of reweighting the data is to reduce the bias or the discrimination in the data. For instance, let's consider a dataset D with a single protected feature S. The advantaged class is denoted by S_a and the disadvantaged class by S_d. Also, the favourable outcome is denoted by Y^+ and the unfavourable outcome by Y^-. We can use statistical parity test or disparate impact for quick insight into the change. But a quicker way would be to calculate discrimination by assigning the weights using the protected class S and the output variable Y.

By carefully choosing the weights (reweighting applies to one protected feature at a time), the training dataset can be made discrimination-free with respect to S without having to change any of the labels or any of the data. The weights act as an additional parameter and can be used directly in any method based on the frequency counts. For instance, let's consider a dataset D with single protected feature S. As discussed in previous chapters, both (Y and S) are binary in nature. In such a case, $X(S) = S_d$ and $Y = Y^+$, disadvantaged class with favourable outcome, will get higher weights than objects with $X(S) = S_a$ and $Y = Y^-$, advantaged class with unfavourable outcome. This is to counter the bias in the data against the S_d and to maintain/increase the accuracy of the positive class. The weights assigned can be altered based on the case by changing the definition of S_a and S_d.

The dataset $(D_{N \times M} = X_i, S_i, Y_i)$ are assigned weights using reweighting technique:

The discrimination, as calculated using the statistical parity difference, before reweighting is

$$P\left(S = S_a \mid Y = Y^+\right) - P\left(S = S_d \mid Y = Y^+\right)$$

With the weights, the discrimination can be calculated using

$$\left\{P\left(S = S_a \mid Y = Y^+\right) \times W_{S_a \wedge Y_{fav}}\right\} - \left\{P\left(S = S_d \mid Y = Y^+\right) \times W_{S_d \wedge Y_{fav}}\right\}$$

where $W_{S_a \wedge Y_{fav}}$ & $W_{S_d \wedge Y_{fav}}$ are the weights applied to the advantageous and favourable and disadvantaged and favourable class.

There will be four categories of weights obtained from the above-stated method, each weight corresponding to the below combination of protected feature (S_a, S_d) and labelled feature (Y^+, Y^- or Y_{fav}, Y_{unfav}):

1. $S_a \wedge Y_{fav}$: S = advantageous (S_a), Y = positive (Y^+ or Y_{fav})
2. $S_a \wedge Y_{unfav}$: S = advantageous (S_a), Y = negative (Y^- or Y_{unfav})
3. $S_d \wedge Y_{fav}$: S = disadvantageous (S_d), Y = positive (Y^+ or Y_{fav})
4. $S_d \wedge Y_{unfav}$: S = disadvantageous (S_d), Y = negative (Y^- or Y_{unfav})

Let's look at how we calculate the weights.

Calculating Weights

The weights are driven by the ratio of expected probability of having favourable outcome to actual probability in the data. In this context and to illustrate the maths behind the method, the protected feature S is defined to be the binary feature that indicates an individual's Marital Status as 0 (not married) or 1 (married), giving us S_a & S_d, respectively. Since not defaulting on a loan is the desired outcome, 0 is the favoured outcome or Y_{fav} and 1 is the Y_{unfav}.

To understand the intuition behind the weights, let's consider unprivileged class and favourable outcome. Using Fig. 5.1, the numbers we will need to be able to understand the intuition are the following:

Total number of observations (n) = 61,321
Total number of observations including unprivileged class (S_d) = 6296
Total number of favourable observations (Y_{fav}) = 46,752
Total number of observations where unprivileged class had a favourable outcome
   ($S_d \wedge Y_{fav}$) = 4043

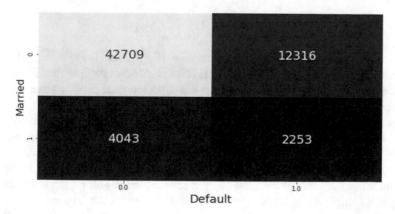

Fig. 5.1 Heat map for feature Married showing a cross tab between the protected feature and the outcome feature

Based on this, the observed probability of unprivileged class having a favourable outcome can be written as ratio of total observations where unprivileged class had a favourable outcome and the total observations.

$$P\left(\text{Observed}_{S_d \wedge Y_{fav}}\right) = \frac{\left(S_d \wedge Y_{fav}\right)}{n}$$

$$\Rightarrow P\left(\text{Observed}_{S_d \wedge Y_{fav}}\right) = \frac{4043}{61321} = 6.59\%$$

Similarly, the expected probability of a person from the unprivileged class having a favourable outcome can be defined as the product of the probability of the person being in the unprivileged group and probability of anyone having a favourable outcome.

$$P\left(\text{Expected}_{S_d \wedge Y_{fav}}\right) = \frac{Y_{fav}}{n} \times \frac{S_d}{n}$$

$$\Rightarrow P\left(\text{Expected}_{S_d \wedge Y_{fav}}\right) = \frac{46752}{61321} \times \frac{6296}{61321} = 7.28\%$$

If the expected probability for a class, given an outcome, is higher than the corresponding observed probability, then the class can be considered to be unprivileged. This must, of course, be then compared with odds for the other classes to compare them and determine which class is the more unprivileged or privileged. In our example, the Marital Status = Married is clearly unprivileged as the expected probability of favourable outcome is 18.73% higher than the observed rate.

A similar calculation for the privileged class and favourable outcome gives us the following numbers.

$$P\left(\text{Observed}_{S_a \wedge Y_{fav}}\right) = 69.65\%$$

$$P\left(\text{Expected}_{S_a \wedge Y_{fav}}\right) = 68.41\%$$

For the privileged class, Marital Status = Single, not only is the difference between expected and observed rate for a favourable outcome less than 2%, but the observed rate is also actually higher than the expected rate! It means that there is a bias acting in favour of the privileged class.

By now applying weights to the records based on the protected feature (in this case Marital Status) and the observed outcome (Y) in the record, we are going to try and bring the observed and expected rates closer to each other. The weights are calculated as a ratio of probability (expected vs observed) of having a favourable outcome. For instance, to calculate weight for privileged group (aka advantageous group) (S_a) and favourable outcome (Y_{fav}), the below steps can be used.

$$\text{Weight}_{S_a \wedge Y_{fav}} = \frac{P\left(\text{Expected}_{S_a \wedge Y_{fav}}\right)}{P\left(\text{Observed}_{S_a \wedge Y_{fav}}\right)}$$

As seen above; $P\left(\text{Expected}_{S_a \wedge Y_{fav}}\right) = \dfrac{Y_{fav}}{n} \times \dfrac{S_a}{n}$

As seen above; $P\left(\text{Observed}_{S_a \wedge Y_{fav}}\right) = \dfrac{\left(S_a \wedge Y_{fav}\right)}{n}$

$$\text{Weight}_{S_a \wedge Y_{fav}} = \dfrac{Y_{fav} \times S_a}{\left(S_a \wedge Y_{fav}\right) \times n}$$

Doing this for all the four categories gives us the following weights, as shown in Table 5.1.

Here the record with $Y = Y_{fav}$ and $S = S_d$ gets a higher weight, and hence an error for this object becomes more expensive (since the higher weight will amplify the effect of the error). This is because we would like to penalize our model most when it makes mistakes on the disadvantageous and favourable group. Similarly, the record with $Y = Y_{unfav}$ and $S = S_d$ gets the lowest weight.

> **Note: For Business Analysts and Product Owners**
> Reweighting the data is not only a highly effective way but also the first opportunity to intervene and fix the bias in the data. The best part about reweighting is that you don't alter the data itself and you do it before even choosing the algorithm for the model. This makes it agnostic of the training algorithm and a highly explainable technique to reduce the bias.
>
> It is however limited in its ability to handle multiple protected features. As we will see later in this chapter, you can overcome that by defining composite features by losing some ability to explain.

Table 5.1 Weights for various combinations of the outcome and the protected feature

$\text{Weight}_{S_a \wedge Y_{fav}}$	0.98227
$\text{Weight}_{S_d \wedge Y_{fav}}$	1.18728
$\text{Weight}_{S_a \wedge Y_{unfav}}$	1.06148
$\text{Weight}_{S_d \wedge Y_{unfav}}$	0.66393

Implementing Weights in ML Model

The weights calculated can be used for most of the commonly used ML algorithms, and most of the standard libraries support weights as part of the data out of the box. These weights would penalize the cost function for any error. In simple words, the algorithm uses weighted errors to penalize based on the weights assigned.

If we have a generic learning algorithm, increasing the sample weight should increase the effect of training on the sample, e.g. if we have a weighted loss, instead of a sum over the training batch, we would do a weighted sum so higher-weighted samples have more impact on the value of the loss. It is as simple as weighing the loss function, such that more important $(S_d \wedge Y_{fav})$ records contribute more strongly to the loss and vice versa.

For instance, the loss function of logistic regression would change to

$$J(\theta) = -\sum_{i=1}^{n} w_i \left[y_i \log p_i + (1 - y_i) \log(1 - p_i) \right]$$

Where w_i is the weight for the record X_i.

And in a tree-based model, weight would impact the impurity measure, nodes split and min weight for leaf node. In case of SVM algorithm, the weight would impact the loss function as

$$J(\theta) = \sum_{i=1}^{n} w_i \times \max\left(0, 1 - y_i f(x_i)\right)$$

Where w_i is the weight for the record X_i.

Let's go through the code to calculate and assign the weights in the data. In the code below, we begin by defining the favourable and unfavourable outcomes and disadvantageous and advantageous classes and create objects that contain the count of each groups.

```
dummy = np.repeat(1, len(data))
data['dummy'] = dummy
n = np.sum(data['dummy']) #Total number of instances
```

```
sa = np.sum(data['dummy'][data[choice]==pval]) #Total number of privileged
sd = np.sum(data['dummy'][data[choice]==upval]) #Total number of unprivileged
ypos = np.sum(data['dummy'][data[target_feature]==fav]) #Total number of favourable
yneg = np.sum(data['dummy'][data[target_feature]==unfav]) #Total number of unfavourable
```

```
print("Total Advantegous: {}, Total Disdvantegous: {}, Total Favourable: {}, Total Unavourable: {}"
    .format(sa, sd, ypos, yneg))
```

```
Total Advantegous: 60895, Total Disdvantegous: 426, Total Favourable: 46752, Total Unavourable: 14569
```

Let's calculate the number of records belonging to the four groups for the combination of two classes and the two possible outcomes.

```
data_sa_ypos = data[(data[choice]==pval) & (data[target_feature]==fav)] # priviliged and favourable
data_sa_yneg = data[(data[choice]==pval) & (data[target_feature]==unfav)] # priviliged and unfavourable
data_sd_ypos = data[(data[choice]==upval) & (data[target_feature]==fav)] # unpriviliged and favourable
data_sd_yneg = data[(data[choice]==upval) & (data[target_feature]==unfav)] # unpriviliged and unfavourable
```

```
sa_ypos = np.sum(data_sa_ypos['dummy']) #Total number of privileged and favourable
sa_yneg = np.sum(data_sa_yneg['dummy']) #Total number of privileged and unfavourable
sd_ypos = np.sum(data_sd_ypos['dummy']) #Total number of unprivileged and favourable
sd_yneg = np.sum(data_sd_yneg['dummy']) #Total number of unprivileged and unfavourable
```

```
print("Total number of the Advantaged and Favourable Group: {}".format(sa_ypos))
print("Total number of Advantaged and Unfavourable Group: {}".format(sa_yneg))
print("Total number of Disadvantaged and Favourable Group: {}".format(sd_ypos))
print("Total number of Disadvantaged and Unfavourable Group: {}".format(sd_yneg))
```

```
Total number of the Advantaged and Favourable Group: 46512
Total number of Advantaged and Unfavourable Group: 14383
Total number of Disadvantaged and Favourable Group: 240
Total number of Disadvantaged and Unfavourable Group: 186
```

Then we calculate the weights. Remember that the formula for the weights (for $S = S_a$ & $Y = Y_{fav}$) is

$$\text{Weight}_{S_a \wedge Y_{fav}} = \frac{Y_{fav} \times S_a}{\left(S_a \wedge Y_{fav}\right) \times n}$$

The formulas for calculating the weights for other combinations follow the same structure.

```
w_sa_ypos= (ypos*sa) / (n*sa_ypos) #weight for privileged and favourable
w_sa_yneg = (yneg*sa) / (n*sa_yneg) #weight for privileged and unfavourable
w_sd_ypos = (ypos*sd) / (n*sd_ypos) #weight for unprivileged and favourable
w_sd_yneg = (yneg*sd) / (n*sd_yneg) #weight for unprivileged and unfavourable

print("Weights for the Advantaged and Favourable Group: {}".format(w_sa_ypos))
print("Weights for the Advantaged and Unfavourable Group: {}".format(w_sa_yneg))
print("Weights for the Disadvantaged and Favourable Group: {}".format(w_sd_ypos))
print("Weights for the Disadvantaged and Unfavourable Group: {}".format(w_sd_yneg))
```

```
Weights for the Advantaged and Favourable Group: 0.9981770630986858
Weights for the Advantaged and Unfavourable Group: 1.0058950456201021
Weights for the Disadvantaged and Favourable Group: 1.3532851714746987
Weights for the Disadvantaged and Unfavourable Group: 0.5441481658390984
```

In the following step, statistical parity difference and weighted statistical parity difference is calculated to judge the overall impact of weights calculated above, and then weights are assigned to respective instances. Statistical parity difference as a measure of discrimination was covered in Chap. 3.

```
datatest=data #.copy()

DiscriminationBefore=(sa_ypos/sa)-(sd_ypos/sd)
DiscriminationAfter=(sa_ypos/sa * w_sa_ypos)-(sd_ypos/sd * w_sd_ypos)

print("Discrimination Before: {}, Discrimination After: {}".format(abs(DiscriminationBefore),
                                                                    abs(DiscriminationAfter)))
```

```
Discrimination Before: 0.20042627057194962, Discrimination After: 0.0
```

```
datatest['Weights']= np.repeat(999, len(datatest))
datatest.loc[(datatest[choice]==pval) & (datatest[target_feature]==fav), 'Weights'] = w_sa_ypos
datatest.loc[(datatest[choice]==pval) & (datatest[target_feature]==unfav), 'Weights'] = w_sa_yneg
datatest.loc[(datatest[choice]==upval) & (datatest[target_feature]==fav), 'Weights'] = w_sd_ypos
datatest.loc[(datatest[choice]==upval) & (datatest[target_feature]==unfav), 'Weights'] = w_sd_yneg
datatest['Weights'].head()
```

```
0    0.998177
1    0.998177
2    0.998177
3    0.998177
4    0.998177
Name: Weights, dtype: float64
```

As expected, after calculating weights in the example above, we can see that the highest weights have been assigned to the $(S_d \wedge Y_{fav})$ group as was our objective.

Using the formula described earlier, the discrimination without using weights was 0.2 while discrimination after using weights is 0.0. In this example, we have a brilliant result; however, the discrimination after using weights won't always reduce to 0. However, it will reduce as compared to the discrimination without weights, and the business analysts in the team will be responsible for prioritizing the features for bias reduction.

In our example, the weights calculated using the above-mentioned concept for a binary protected feature are used as sample weight for logistic regression with the response variable being binary feature and all other features are as-is. The data was split into test and train data with 30:70. In the first model, all variables were included without any sample weight, while for the second model, the sample weights as calculated above were included.

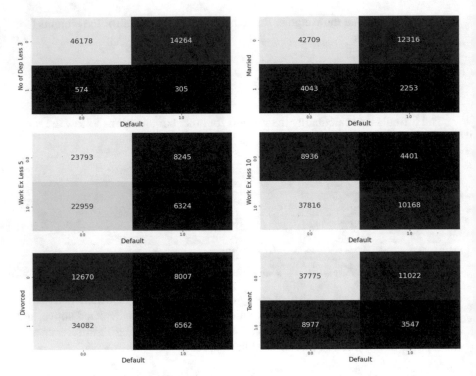

Fig. 5.2 Heat map showing relationship of six protected features with outcome feature

One of the key concerns that the product owner and the business analysts may have at this stage are the impact of this technique on the accuracy of the model. As we will see in the examples below, the overall accuracy of models with weights improves by a few points, while the overall accuracy difference between privileged group and unprivileged group has become narrower. A win-win situation.

For the data in discussion, let us look how reweighting impacted the discrimination and was it able to make the process and the model relatively fairer. We will first look at the three groups of records based on the Marital Status (Married, Single and Divorced). These three features have been built by applying one hot encoding on a single Marital Status feature. Looking at the heat maps, in Fig. 5.2, for these features vis-à-vis the default prediction, we can see that the groups Married and Single are unprivileged as compared to the group Divorced.

Protected Feature: Married

We can see in Table 5.2 that the maximum weight after reweighting was given to the unprivileged class and favourable outcome, again proving that we wanted to ensure more fairer decision for the given combination.

The charts in Fig. 5.3 are quite interesting, for the group Marital Status = Married, we ran a logistic regression model without weights and with weights (as computed

Table 5.2 Weights for the protected feature: Married

	$S_a \wedge Y_{fav}$	$S_a \wedge Y_{unfav}$	$S_d \wedge Y_{fav}$	$S_d \wedge Y_{unfav}$
Weights	0.98227	1.06147	1.18727	0.66393

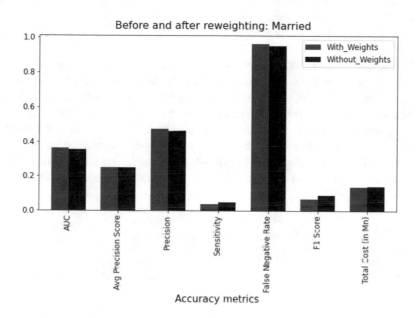

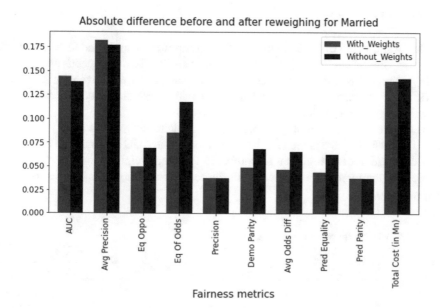

Fig. 5.3 Plots for model performance before and after reweighting; absolute difference in fairness metric for protected feature

by reweighting algorithm). It is clear that even after adding weights, the overall AUC, false negative rate (FNR) and precision didn't dip much while there was some decrease in sensitivity and F1 score. Depending on the use case and the improvement achieved, the business may zero-in on a few metrics on which they would like to judge the overall model. The product owner and the business analysts have an important decision to make when it comes to deciding which metrics (like AUC, FNR or FPR, F1, precision or recall) need to be optimized. See the end of this chapter for a note on the impact of reducing false positive and false negative rates.

We can also see in Table 5.3 that reweighting technique decreases the discrimination (using statistical parity difference) to zero. But here, the second plot is what we are interested in. The plot on the right shows that the important fairness metric like equalized odds, equal opportunity and demographic parity has gone fairer (reduced) as compared to what it was pre-reweighting logistic regression. A few secondary metrics also yield a fairer out with weights in the model. Again, we would like to emphasize that a PO or BA needs to prioritize the metric that needs to be focussed on. See the note at the end of the next chapter for an example prioritization of these metrics.

Table 5.3 Reduction in discrimination after applying reweighting

Reweighting	Before	After
Discrimination (SPD)	0.1340	1.11022e-16

Protected Feature: Single

Table 5.4 shows the weights assigned to four groups for the protected feature Marital Status = "Single". In Fig. 5.4, we can see that as above, disadvantageous and favourable group got highest weight. Here the results are similar as above with AUC and FNR showing improvement in reweighted logistic regression model compared to baseline logistic regression model.

Here, we can see in Table 5.5 that the discrimination has been brought to zero. But what is better here is the fact that even metric like AUC and predictive equality saw decrease in discrimination as compared to a non-weighted logistic regression model. Furthermore, all the three main fairness metric illustrates comparatively a fair outcome.

Table 5.4 Weights for protected feature: Single

	$S_a \wedge Y_{fav}$	$S_a \wedge Y_{unfav}$	$S_d \wedge Y_{fav}$	$S_d \wedge Y_{unfav}$
Weights	0.97317	1.09703	1.30054	0.57419

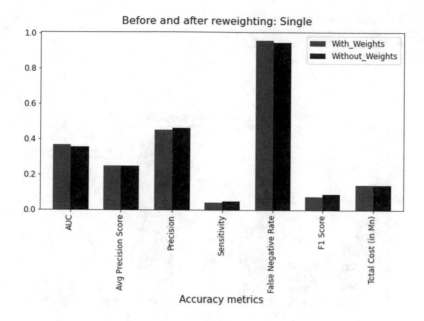

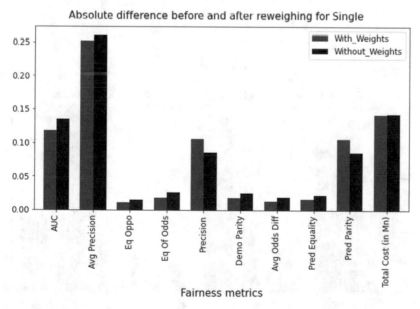

Fig. 5.4 Plots for model performance before and after reweighting; absolute difference in fairness metric for protected feature: Single

Protected Feature: Divorced

Finally, for the third group in Marital Status, we see in Fig. 5.5 that the weights don't follow the same pattern as before. The highest weight is assigned to $S_a \wedge Y_{unfav}$, which represents the records with Marital Status = Divorced and Outcome = Default.

Table 5.5 Reduction in discrimination after applying reweighting

Reweighting	Before	After
Discrimination	0.1972	0

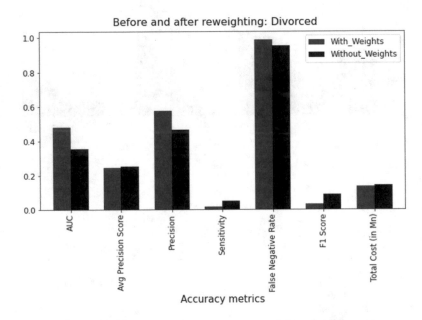

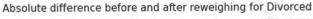

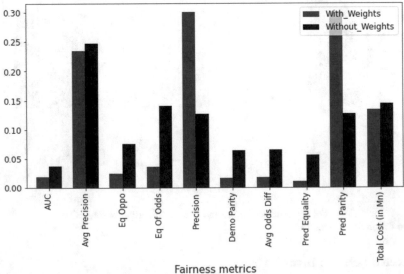

Fig. 5.5 Plots for model performance before and after reweighting; absolute difference in fairness metric for protected feature: Divorced

Table 5.6 Weights for the protected feature: Divorced

	$S_a \wedge Y_{fav}$	$S_a \wedge Y_{unfav}$	$S_d \wedge Y_{fav}$	$S_d \wedge Y_{unfav}$
Weights	0.90920	1.47156	1.24423	0.61353

Table 5.7 Reduction in discrimination after applying reweighting

Reweighting	Before	After
Discrimination	0.2257	1.1102e-16

However, for the records representing the unprivileged class, the higher weight is still assigned to the $S_d \wedge Y_{fav}$ as expected.

To avoid higher weights for $S_a \wedge Y_{unfav}$ group, we recommend generating weights for an unprivileged class instead of the privileged class. The weights for the four groups are shown in Table 5.6.

By generating weights for the privileged class, we can see that even though the discrimination goes down, as shown in Table 5.7, it is at a significant cost to the precision and the predictive parity.

Protected Feature: Number of Dependants Less than Three

Now we will look at two more protected features starting with this one. For this feature, the privileged group is the one with less than three dependants, and the unprivileged group is the one with three or more dependants. The weights calculated are shown in Table 5.8; we can see that the maximum weight got assigned to disadvantageous and favourable group. Even here, in Fig. 5.6, we see no dip in overall AUC or other model accuracy metric.

As expected, the discrimination using statistical parity difference measure is zero (see Table 5.9), while we see that AUC, precision and predictive parity have worsen off after reweighting treatment, but there is a significant improvement in fairness if measured against fairness metric like equal opportunity and demographic parity. It is evident that all metrics didn't get optimized, with few showing considerable improvement while others get worse off. This once again highlights the importance of prioritizing these metrics from a business point of view by the PO and the BAs in the team.

Table 5.8 Weights for the protected feature: Number of dependants less than three

	$S_a \wedge Y_{fav}$	$S_a \wedge Y_{unfav}$	$S_d \wedge Y_{fav}$	$S_d \wedge Y_{unfav}$
Weights	0.99817	1.00589	1.35328	0.54414

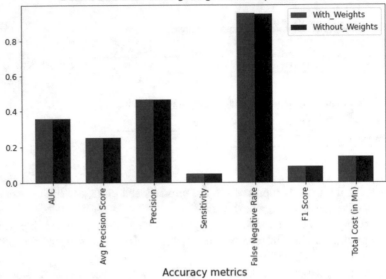

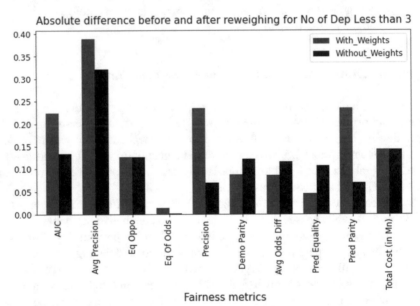

Fig. 5.6 Plots for model performance before and after reweighting; absolute difference in fairness metric for protected feature: No. of dependants less than three

Table 5.9 Reduction in discrimination after applying reweighting

Reweighting	Before	After
Discrimination	0.2004	0

Protected Feature: Work Experience Less than 10 Years

In the case of protected feature Work Experience less than 10 years, it follows the first example with disadvantageous and favourable group receiving the maximum weight, as shown in Table 5.10, and a slight improvement in overall model accuracy, if measured using AUC, compared to a non-weighted logistic regression model, as shown in Fig. 5.7.

Table 5.10 Weights for the protected feature: Work experience less than 10 years

	$S_a \wedge Y_{fav}$	$S_a \wedge Y_{unfav}$	$S_d \wedge Y_{fav}$	$S_d \wedge Y_{unfav}$
Weights	0.96741	1.12119	1.13790	0.71999

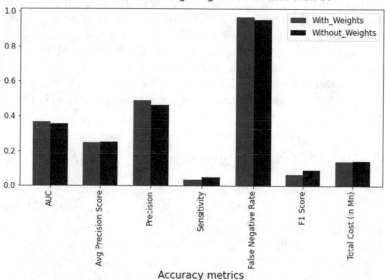

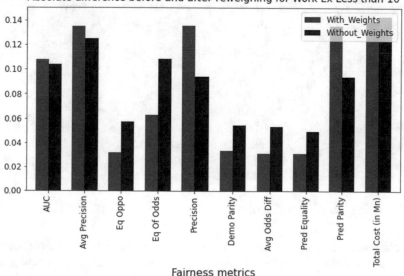

Fig. 5.7 Plots for model performance before and after reweighting; absolute difference in fairness metric for protected feature: Work experience less than 10 years

Table 5.11 Reduction in discrimination after applying reweighting

Reweighting	Before	After
Discrimination	0.11808	0

The discrimination here is reduced to 0, as shown in Table 5.11, and most of the fairness metrics also show significant improvement, except precision and predictive parity.

After using weights (generated from reweighing algorithm), the values of metrics like AUC, precision and false negative rate have increased compared to model where weights were not used, while sensitivity, F1 score and cost have decreased significantly. However, what we observe is that the total cost of the model has decreased slightly. The cost here is calculated as Total Cost = 700*FP + 300*FN (given the fact that in this case, false positive or FP would be extremely costly for any lender compared to false negative or FN – see the note at the end of this chapter for more details.)

Note: Quick Summary of Reweighting Approach

As we discussed at the beginning of this chapter, reweighting can only be applied on one feature at a time. The steps we took are the following:

- Choose a protected feature to generate weights.
- Generate weights and add them to the dataset.
- Choose an algorithm to train a model using both unweighted and weighted data.
- Calculate statistical parity difference for both models to compare the discrimination.
- Generate the accuracy and fairness metrics to compare the difference between the two group and between two models (with and without weights).
- Repeat the above steps for all the protected features you want to analyse and compare to decide.

Based on your business problem and the results of the analysis above, you'd want to choose a feature to generate weights and reduce discrimination because of it or combine multiple features into a single composite feature if that helps achieve better reduction in overall discrimination while achieving acceptable performance across other key metrics as defined.

Calibrating Decision Boundary

The classification algorithms we have used so far generate a probability of the loan record in the dataset defaulting on the payment. This is converted into a binary decision by applying a decision threshold. For our examples, the decision threshold is

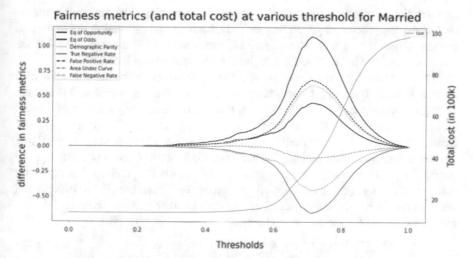

Fig. 5.8 Plot showing the value of various fairness metric across all threshold values

0.5 – No Default <0.5 and Default ≥0.5. Before proceeding further, let's try to understand the impact of threshold on the fairness metrics.

The charts in Fig. 5.8 show how each fairness metric and the total cost changes as we change the decision threshold from 0 to 1 for the protected feature Marital Status = Married. A decision threshold of 0 means that every record gets a negative outcome (Default), whereas a decision threshold of 1 means that every record gets a positive outcome (No Default). Therefore, even though difference in all fairness metrics seem to converge close to 0, as we get close to the two extremes for the threshold, it is not viable business strategy to mark every record as likely to default, likely resulting in no loan being granted or every record being not likely to default, resulting in everyone who applied getting a loan – clearly neither of these two options will make for a sound lending strategy!

By changing the decision threshold, we can impact the fairness metrics. Therefore, the balance therefore lies somewhere in between – a threshold that achieves the best fairness outcome possible while minimizing the cost function.

We can see that for protected feature Marital Status = Married, the cost curve has a point of inflection at about threshold = 0.6 and the fairness metrics (difference between two groups of the protected class) seem to be minimum for all measures at threshold ≤0.5. Thus, it would be optimal to choose a decision threshold such that 0.5 ≤ threshold ≤0.6.

Composite Feature

When you have multiple protected features, it is not possible to mitigate bias for each one of them individually. In such cases, it may make sense to use a logical operator to create composite protected feature. A composite feature can help reduce bias due to multiple features at the same time. Once you have defined your composite feature, the process of generating weights, training and estimating the impact on discrimination (along with accuracy and fairness metrics) is the same as what we have used so far. The real trick lies in defining the composite feature.

For our example problem, we may want to combine one of the marital status with another feature. The marital status has three groups: Married, Single and Divorced. Of these three, for our data and problem, the Married and Single are unprivileged, and the Divorced is the privileged class. Therefore, we can use Marital Status! = Divorced (i.e. "MaritalStatus_4.0" = 0) as one of the features to combine. The other feature we are picking for our example is Work Experience less than 10 years. The group with experience of less than 10 years is the privileged group in our dataset; hence, we pick up the unprivileged group (or "WrExLess10" = 0). Note that we are combining one unprivileged group with another and not with a privileged group. We define the composite feature as

data['Combined_protected_group'] = np.where((data['WrExLess10'] == 0) & (data['MaritalStatus_4.0'] == 0),0, 1)

This composite feature gives us the heat map and the weights as shown in Fig. 5.9, and the associated weights for the four groups are shown in Table 5.12.

It is evident from heat map shown in Fig. 5.9 that the disadvantageous group definition is mathematically sane.

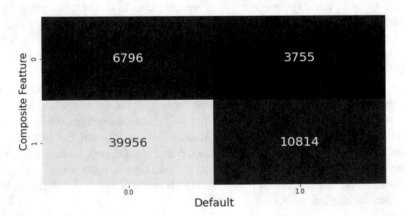

Fig. 5.9 Heat map representing cross tab between composite feature and outcome feature

Table 5.12 Weights for the composite feature

	$S_a \wedge Y_{fav}$	$S_a \wedge Y_{unfav}$	$S_d \wedge Y_{fav}$	$S_d \wedge Y_{unfav}$
Weights	0.9687	1.1154	**1.1836**	0.6675

Training a logistic regression model on this gives us the output shown in Fig. 5.10. We can see that this also reduces the statistical parity difference significantly, as shown in Table 5.13.

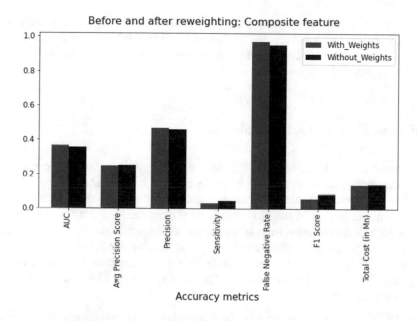

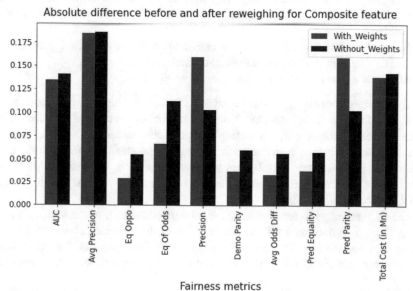

Fig. 5.10 Plots for model performance before and after reweighting; absolute difference in fairness metric for composite protected feature

Table 5.13 Reduction in discrimination after applying reweighting

Reweighting	Before	After
Discrimination	0.1428	0

Composite feature can be a very powerful concept when engineering cautiously. This allows not only to overcome the issues of reweighing (of handling multipole protected features) but also would allow to combine protected features where the disadvantageous representation is very low.

Additive Counterfactual Fairness

Now that we have looked at reweighting and its capabilities, let's explore another method which takes a very different approach. The fundamental principle behind additive counterfactual fairness is related to causality and is quite simple – it aims to take out the impact of the protected feature from the model if it influences the output in a discriminatory way. It states that a protected feature should not be a cause for the predictor to be discriminatory – or that the predicted outcome should not change if the value of the protected feature is changed from a privileged group to an unprivileged group or vice versa. The key features of the additive counterfactual approach are the following:

1. ACF (additive counterfactually fair) models can be implemented using any machine learning algorithm and are apt for most of regression and classification problems.
2. No need for composite features as it can tackle multiple protected features.
3. Can handle continuous and categorical protected features.
4. ACF may cause a small dip in the accuracy.

The primary concept of ACF is based on causality. A causal graph is counterfactually fair if the predicted outcome $\hat{Y}$ in the graph does not depend on a descendant of the protected attribute S. For example, a predictive outcome $\hat{Y}$ of defaulter vs non-defaulter for a loan application is typically dependent on credit score, credit amount, disposable income and years of work experience. For instance, if disposable income is a direct descendant of number of dependants, the given causal model is not counterfactually fair.

The ACF attempts to remove these causal dependencies through explicitly modelling all input variables as a linear combination of the protected class variables. By taking the residuals as a difference between actual (X) and predicted input variables ($\hat{X}$), we can effectively remove the correlation between the protected classes to the input variables.

ACF, within the scaffold of counterfactual fairness, is the concept of modelling the relationship between S and features in X by training additive models to predict each feature X_j (as the outcome feature) with S as the predictors.

Then, we can compute the residuals ϵ_{ij} between predicted values ($X - \hat{X}$) and true feature values (X) for each observation i and non-protected feature X_j. The final model is then trained on the residuals (ϵ_{ij}) as features to predict the outcome feature Y.

High Level Steps for Implementing ACF Model

1. Develop a separate model to predict each of the independent features (non-protected) using protected features as the predictor features.
2. Compute the residuals for each independent feature.
3. Develop a model with Y as response and residuals as predictors.

$$\widehat{X_1} = f_1\left(S_1, S_2, \ldots, S_n\right)$$

$$\vdots$$

$$\widehat{X_n} = f_n\left(S_1, S_2, \ldots, S_n\right)$$

$$\epsilon_{X1} = X_1 - \widehat{X_1}$$

$$\vdots$$

$$\epsilon_{Xn} = X_n - \widehat{X_n}$$

$$\hat{Y} = f_y\left(\epsilon_{X1}, \epsilon_{X2}, \ldots, \epsilon_{Xn}\right)$$

The dataset used here is the subset of the data previously used. For ease of illustration, we will use only a subset of the dataset and not entire dataset.

The choice of using ACF model over other techniques is because ACF allows lot of flexibility and is model agnostic and has simple implementation without a need of creating new algorithm. It uses the baseline model (which can be any supervised learning model) and models on the top of the baseline model.

Now let's look in more detail at how ACF can be applied on two kinds of problems (classification and regression) using our sample dataset.

ACF for Classification Problems

Let's continue with the example we are discussing. Here again we want to develop a model that classifies customers as defaulter and non-defaulter. For ease, we are using a small subset of independent features.

- Here the target variable is:
  - Default

- The independent variables are:

  - Age
  - Applied amount
  - Expected loss
  - Liabilities total
  - Income total
  - Interest
  - Loan duration
  - Monthly payment

- The sensitive variables are:

  - Number of dependants less than three
  - Married
  - Single
  - Divorced

In the code below, we are initially declaring the protected features and then creating eight independent models with the protected features as predictors and non-protected features as response feature. After that, we calculate the residuals for each of the models, and the residuals are then used as predictor for a final ACF model and Y (default) as response.

Please note that, here we are not tuning the model and are simply using a linear model just for illustration. It is up to the developers to tune model at any stage and use any linear/nonlinear model. If required, in order to improve the model accuracy, one can swap the linear or logistic regression model with a tuned version or complex model.

Selecting few protected / sensitive features

```
sens=X_train[["NrOfDependantslessthan3", 'MaritalStatus_1.0', 'MaritalStatus_3.0', 'MaritalStatus_4.0']]
```

Regressing each independent variable with sensitive variable

```
clf_age = LinearRegression().fit(sens, X_train['Age'])
clf_AppliedAmount = LinearRegression().fit(sens, X_train['AppliedAmount'])
clf_ExpectedLoss = LinearRegression().fit(sens, X_train['ExpectedLoss'])
clf_LiabilitiesTotal = LinearRegression().fit(sens, X_train['LiabilitiesTotal'])
clf_IncomeTotal = LinearRegression().fit(sens, X_train['IncomeTotal'])
clf_Interest = LinearRegression().fit(sens, X_train['Interest'])
clf_LoanDuration = LinearRegression().fit(sens, X_train['LoanDuration'])
clf_MonthlyPayment = LinearRegression().fit(sens, X_train['MonthlyPayment'])
```

Residual of each model

```
ageR = X_train['Age'] - clf_age.predict(sens)
AppliedAmountR = X_train['AppliedAmount'] - clf_AppliedAmount.predict(sens)
ExpectedLossR = X_train['ExpectedLoss'] - clf_ExpectedLoss.predict(sens)
LiabilitiesTotalR = X_train['LiabilitiesTotal'] - clf_LiabilitiesTotal.predict(sens)
IncomeTotalR = X_train['IncomeTotal'] - clf_IncomeTotal.predict(sens)
InterestR = X_train['Interest'] - clf_Interest.predict(sens)
LoanDurationR = X_train['LoanDuration'] - clf_LoanDuration.predict(sens)
MonthlyPaymentR = X_train['MonthlyPayment'] - clf_MonthlyPayment.predict(sens)
```

```
df_R=pd.DataFrame({'ageR':ageR, 'AppliedAmountR':AppliedAmountR, 'ExpectedLossR':ExpectedLossR,
                   'LiabilitiesTotalR':LiabilitiesTotalR,
                   'IncomeTotalR':IncomeTotalR, 'InterestR':InterestR, 'LoanDurationR':LoanDurationR,
                   'MonthlyPaymentR':MonthlyPaymentR})
```

Fitting residuals to the dependent (target) variable

```
fair = LogisticRegression(random_state=0, solver='liblinear',
                          multi_class='ovr').fit(df_R, y_train)
```

Table 5.14 AUC scores for logistic regression vs ACF models

	AUC scores
Logistic regression model	0.68808
ACF model	0.60861

After training the model, we can perform the prediction or get the prediction probability of the logistic regression model with Y (default) as the response feature for calculating model accuracy.

Predicting the target variable on fair model developed above

```
pred_fair_te = fair.predict(df_R_test)
```

```
pred_fair_te_prob=fair.predict_proba(df_R_test)[:,0]
```

Please note the results shown in Table 5.14, defined with AUC metric here, where we see a few points dip in AUC of ACF model compared to AUC of a logistic regression model. As stated earlier, please note that the regression algorithm illustrated here can be replaced with any algorithm of your choice.

Table 5.15 shows how the same ACF models fairs for different protected features. The AUC difference here is computed as difference between roc value of the prediction with advantageous group and disadvantageous group of the protected features. In almost all the cases, we see that the fairness metric is performing significantly better for all protected feature though the AUC seems to have dipped a bit. But that was the trade-off we mentioned before. However, the trade-off can be reduced by the following:

1. Optimizing decision boundary
2. Selecting better set of protected features
3. Using tuned or a complex ML model

Now let's use ACF for a continuous response feature.

ACF for Continuous Output

For regression-based illustration, we would use the same subset of the dataset as previously used with a change in the response feature. Instead of Y being a binary feature representing default, we would replace the Y with a continuous feature Default Score (values ranging between 0 and 850)

Table 5.15 Difference realized in fairness metrics after applying ACF

Feature	Absolute difference in fairness metrics and total cost of the model
No. of dependants less than three	
Divorced	
Single	

Absolute difference (and total cost): full vs ACF model No of Dep Less 3

Absolute difference (and total cost): full vs ACF model for Divorced

Absolute difference (and total cost): full vs ACF model for Single

Table 5.15 (continued)

Feature	Absolute difference in fairness metrics and total cost of the model
Married	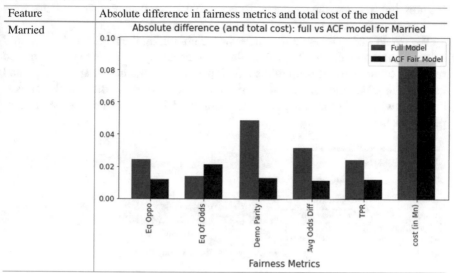

The target variable (Y) is Default Score (ranges between 0 – 850).
The independent variables (X) are same as above:

- Age
- Applied amount
- Expected loss
- Liabilities total
- Income total
- Interest
- Loan duration
- Monthly payment

The sensitive variables (S) are as discussed earlier:

- Number of dependants less than three
- Married
- Single
- Divorced

Here for comparison, two models were created one being a full-linear regression model (M_{Lin}) to predict the default score and another a linear regression model using ACF approach (with eight underlying models to generate the residuals).

Here for evaluation instead of fairness metrics, the distribution of prediction across various protected groups was used.

In the code below, we declare the protected features and then create eight independent models with the protected features as predictors and non-protected features as response feature. Once we have our models, we use them to predict independent features and calculate the residuals. These residuals are then used as predictor for a final ACF model as predictor and Y (default) as response.

We would like to reiterate that here we are not tuning the model and are simply using a linear model for illustration; thus, the accuracy of the model may not be really up to the mark. It is up to the developers to tune the model at any stage and use any linear/nonlinear model. If required, in order to improve the model accuracy, one can swap the linear or logistic regression model with a tuned version or complex model.

Selecting few protected / sensitive features

```
sens=X_train[["NrOfDependantslessthan3", 'MaritalStatus_1.0', 'MaritalStatus_3.0', 'MaritalStatus_4.0']]
```

Regressing each independent variable with all sensitive variables

```
clf_age = LinearRegression().fit(sens, X_train['Age'])
clf_AppliedAmount = LinearRegression().fit(sens, X_train['AppliedAmount'])
clf_ExpectedLoss = LinearRegression().fit(sens, X_train['ExpectedLoss'])
clf_LiabilitiesTotal = LinearRegression().fit(sens, X_train['LiabilitiesTotal'])
clf_IncomeTotal = LinearRegression().fit(sens, X_train['IncomeTotal'])
clf_Interest = LinearRegression().fit(sens, X_train['Interest'])
clf_LoanDuration = LinearRegression().fit(sens, X_train['LoanDuration'])
clf_MonthlyPayment = LinearRegression().fit(sens, X_train['MonthlyPayment'])
```

Residual of each models

```
ageR = X_train['Age'] - clf_age.predict(sens)
AppliedAmountR = X_train['AppliedAmount'] - clf_AppliedAmount.predict(sens)
ExpectedLossR = X_train['ExpectedLoss'] - clf_ExpectedLoss.predict(sens)
LiabilitiesTotalR = X_train['LiabilitiesTotal'] - clf_LiabilitiesTotal.predict(sens)
IncomeTotalR = X_train['IncomeTotal'] - clf_IncomeTotal.predict(sens)
InterestR = X_train['Interest'] - clf_Interest.predict(sens)
LoanDurationR = X_train['LoanDuration'] - clf_LoanDuration.predict(sens)
MonthlyPaymentR = X_train['MonthlyPayment'] - clf_MonthlyPayment.predict(sens)
```

```
df_R=pd.DataFrame({'ageR':ageR, 'AppliedAmountR':AppliedAmountR, 'ExpectedLossR':ExpectedLossR,
                   'LiabilitiesTotalR':LiabilitiesTotalR,
                   'IncomeTotalR':IncomeTotalR, 'InterestR':InterestR, 'LoanDurationR':LoanDurationR,
                   'MonthlyPaymentR':MonthlyPaymentR})
```

Fitting residuals to the dependent (target) variable

```
fair = LinearRegression().fit(df_R, y_train)
```

After training the model, let's predict the output Y (default score) as the response feature and a few metrics for model evaluation.

Predicting the target variable on fair model developed above

```
pred_fair_te = fair.predict(df_R_test)
```

```
print("For ACF model:")
print("Mean Squared Error:",mean_squared_error(y_test, pred_fair_te))
print("Root Mean Squared Error:", RMSE(pred_fair_te, y_test))
print("Mean Absolute Percentage Error:", mape(pred_fair_te, y_test))
```

As mentioned initially, we observe a small dip in the accuracy of the ACF model compared to linear regression model as shown in Table 5.16. The trade-off can be altered by using a better or tuned regression model.

Tables 5.17 and 5.18 show the model accuracy difference in case of regression model and ACF models. The difference here is between the model accuracy of advantageous group and disadvantageous group for the mentioned protected feature using the same models.

Table 5.16 Model accuracy parameters for linear regression vs ACF

	Mean squared error	Root mean squared error	Mean absolute Percentage Error
Linear regression	1991.3528	44.6245	22.8620
ACF	2442.6078	49.4227	28.2511

Linear Regression Model

Table 5.17 Model accuracy differences for various protected groups when using linear regression model

Linear regression model	Mean squared error difference	Root mean squared error difference	Mean absolute percentage error difference
Single	517.3058	6.1346	1.0619
Married	873.3461	10.8661	2.0646
Divorced	917.6351	10.7927	0.6545
No of Dependants less than 3	296.3293	3.4520	4.3778

ACF Model

Table 5.18 Model accuracy differences for various protected groups when using ACF model

ACF model	Mean squared error difference	Root mean squared error difference	Mean absolute percentage error difference
Single	840.4945	9.1932	3.6977
Married	1279.3378	11.8464	12.0020
Divorced	130.5891	1.3272	3.1517
No. of Dependants less than three	773.6410	7.2958	7.7870

And finally, as seen in Table 5.19, the mean ratio (and skewness and kurtosis ratio) between advantageous and disadvantageous groups (the two overlapping curves below) of a protected feature seems to improve significantly after ACF model is wrapped on a regression model.

Table 5.19 Density plots and comparison of mean, skewness and kurtosis for the various protected features when using linear regression vs ACF model

Density plots of predictions by protected groups

	Linear regression model	ACF model
Single		
Mean	1.0744	1.0000
Skewness	0.4424	0.6936
Kurtosis	0.4371	0.8704
Married		
Mean	1.3801	1.0000
Skewness	0.5896	0.7765
Kurtosis	0.7095	1.0573

Density plots of predictions by protected groups

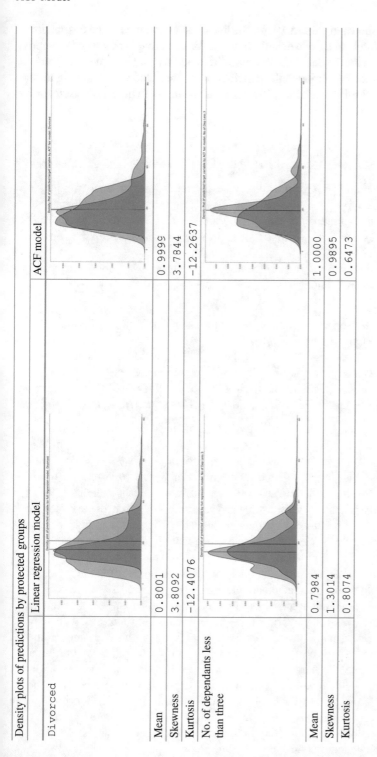

	Linear regression model	ACF model
Divorced		
Mean	0.8001	0.9999
Skewness	3.8092	3.7844
Kurtosis	-12.4076	-12.2637
No. of dependants less than three		
Mean	0.7984	1.0000
Skewness	1.3014	0.9895
Kurtosis	0.8074	0.6473

In the charts shown in Table 5.19, for the linear regression model, the distribution of the target value (probability of a default) between the two groups (privileged vs unprivileged) is quite varied as evident from the large difference in mean, skewness and kurtosis.

However, for the ACF model, the distribution of predicted values among the two groups is very similar, and the difference in mean, skewness and kurtosis has reduced dramatically.

Note: Understanding Residuals and Determining Independence of Features

ACF is an incredibly powerful technique for reducing bias due to multiple features at the same time. It also allows us to determine if the "independent" features are truly independent or if they are dependent on some of the protected features and can be proxies for them – as that would mean they are no longer independent.

To understand how the potential relationship between the independent and protected features is uncovered, let's recap one more time the steps we follow for ACF:

1. We begin by separating our dataset into protected features (S_j) and independent features (X_j).
2. We create a separate model to predict each independent feature using the protected features as the predictor, or inputs. We represent these predictions as $(\widehat{X_j})$.
3. We calculate the residuals as the difference between the actual and predicted values for the independent features. This is represented by ϵ_j.

Now, let's pause here and try to understand the residuals in more detail. The residuals, ϵ_j, are nothing but the error in the prediction of the independent features. And we know that a small value for the error means better prediction using the predictors, which in this case are the protected features. Therefore, if the residuals for a given independent feature have an overall small value, then that independent feature is not truly independent and can serve as a proxy for the protected features. Similarly, if the residual values are high, then that indicates relative independence from the protected features and hence reduced ability of that feature to serve as a proxy for the protected features – which is what we desire!

As you determine the relationships between the protected and the independent features, you need to decide how to handle the features that were considered to be independent but have clear relationship with the protected features. For features with strong relationship with the protected features, you may consider dropping them altogether from your dataset or adding them to the protected features (proxy protected feature) and start again. On the other hand, if the protected features fail to explain an independent feature – indicated by high residual values – you may want to keep the feature out of the treatment and use it as-in in the final model. Going a step further, through multiple iterations, the business analyst and the data scientists working together may pick and choose a custom set of protected features to be regressed with the independent features to find better relationships between the two sets and then decide the modelling approach based on the findings.

Note: Different Types of Residuals

Now that we understand the behaviour of residuals a little better, let's talk about how we can calculate them in different ways and what kind of impact they can have on our final ACF model when it comes to reducing the bias. The residual, ϵ_j, as we discussed in the previous note is the error in predicting the independent feature using protected feature as the predictors.

In this chapter, we have calculated the residual as just the difference between the actual and predicted values for the independent features $\left(\epsilon_j = X_j - \widehat{X}_j\right)$. This can give us a positive or a negative value for the residual, based on the difference. However, as discussed earlier, it is the magnitude of the residual that indicates the ability of the protected features to predict the independent features. Both large positive and negative values for the residual indicate the relative independence of the feature from the protected features.

Armed with this idea, we would like to share few more possible ways to calculate the residuals.

Absolute value: $\epsilon_j = \left\lvert X_j - \widehat{X}_j \right\rvert$

This calculates residual as the absolute value of the difference between the actual and the predicted values for the independent features. For our example, if we use absolute value of the difference as the residual, the discrimination measured for the protected feature "Marital Status = Single" drops to 2.39% from 3.6% we get when calculating the residual as shown so far.

Residual squared: $\epsilon_j = \left(X_j - \widehat{X}_j\right)^2$

Just like the example above, this approach will remove the ± signs from the value of the residual. In addition, this will also amplify the distance of residual from 0 by squaring the value. For our example, if we use squared value of the difference as the residual, the discrimination measured for the protected feature "Marital Status = Single" drops to 2.4% from 3.6%. For our example, these two approaches yield a similar result, but you'd need to decide if you want to explore these as options and include one or more approach in your analysis, since the residuals act as the predictors for the final model.

One of the challenges of using squared value of the difference as the residual is that all the values between 0 and 1 will get compressed into an even smaller range, whereas all values over 1 will get spread apart further. We can overcome that by altering the definition of the residual to

$$\epsilon_j = (M + (X_j - \widehat{X}_j))^2$$

Where $M \gg 1$, basically a large positive integer to make all values higher than 1 and spread the values further apart. For our example, if we choose $M = 100$, the discrimination measured for "Marital Status = Single" drops to 2.28% as opposed to 2.4% when $M = 0$.

In some cases, the independent variables may be binary which would require a classification model to be developed between the independent feature and the sensitive features. In such a scenario, to calculate the residual (which will be one of the predictors for the final model), it's advisable to use either Pearson residual or Deviance residual to calculate the error.

Pearson residual is basically residual normalized by dividing the residual by square root of the estimate.

$$\text{residual} = \frac{y_i - \widehat{y}_i}{\sqrt{\widehat{y}_i\left(1-\widehat{y}_i\right)}}$$

Deviance residual is the difference of the log-likelihoods between the fitted model and the saturated model.

$$\text{residual} = \sqrt{-2\left[y_i\log\widehat{y}_i + \left(1-y_i\right)\log\left(1-\widehat{y}_i\right)\right]}$$

The different approaches to calculating the residual are part of your tool kit as you use ACF to improve the fairness of the predictions. We would suggest treating this step as a part of the exploratory analysis and feature engineering to determine the residual you want to use. In a few cases, you may even opt to have different residual approaches for different independent features, where a few select independent features; residual can be simple residual, while for others it can be squared residual, and may be for a few, one can opt for absolute residuals.

Calculating Unfairness

So far in this section, we have looked at how to develop an ACF model for two types of problems and also how to judge the effectiveness of the model based on the two types of the algorithms. Now let's talk about how to assess if we need an ACF model or if an ACF model is adding any value. To better understand the effectiveness of the ACF model, we are introducing the concept of counterfactual unfairness (CUF). We aim to use CUF to determine if the ACF model we have created is able to remove the impact of sensitive features from decision-making along with improving fairness. The theme of counterfactual unfairness is "What would have been the prediction if – all else held causally constant – the record belonged to a member of another protected group?". For example, would the predicted outcome change from No Default to Default if the record is switched from Divorced (privileged group) to Single (unprivileged group), or would it remain No Default as we would expect?

On a high level, the approach for counterfactual unfairness is as follows:

1. Create the ACF model as described before:

   (a) Train models ($X_{full1}...X_{fulln}$) to predict the independent features with the protected features (S) as predictors.
   (b) Use protected features (S) to predict the independent features ($\widehat{X_j}$).
   (c) Calculate the residuals ($\epsilon_j = X_j - \widehat{X_j}$).
   (d) Train the model (MACF) to predict the output ($\hat{Y}$) with residuals (ϵ_j) as predictors.

2. Generate the predictions using the ACF model created (MACF) ($\hat{Y}$).
3. Calculate the error in prediction ($E = Y - \hat{Y}$).
4. Invert the protected features – change privileged to unprivileged and vice versa (S').
5. Predict independent features ($\widehat{X'}$) using inverted protected features (S') as the input (predictors) to the model ($X_{full1}...X_{fulln}$) from above.
6. Calculate the residuals ($\epsilon_j' = X \ \widehat{X'}$).
7. The residuals (ϵ') act as the input to the ACF model MACF to generate the predictions ($\widehat{Y'}$).
8. Calculate the error in prediction ($E' = Y - \widehat{Y'}$).
9. Calculate the counterfactual unfairness $\left[CUF = \dfrac{1}{n}\Sigma(E - E')^2 \right]$. A smaller value of CUF means that our ACF model has less impact due to sensitive features.

Now to compare this with a model where we have not applied ACF, we can take the following steps:

1. Train any model to predict the output (M_{linreg}).
2. Generate the predictions ($\widehat{Y_{linreg}}$).
3. Calculate the error ($E_{linreg} = Y - \widehat{Y_{linreg}}$).
4. Invert protected features (S').
5. Generate the predictions again ($\widehat{Y'_{linreg}}$).
6. Calculate the error ($E'_{linreg} = Y - \widehat{Y'_{linreg}}$).
7. Calculate CUF $\left[CUF = \dfrac{1}{n}\Sigma\left(E_{linreg} - E'_{linreg}\right)^2 \right]$.

We can now compare the two CUF values to understand the impact of the bias reduction.

The code below demonstrates error calculated for a regression-based model for model with sensitive features (S) and counterfactual sensitive features (S'). The CUF is then calculated as difference of both the error.

The first step was to develop a full logistic model and note the error of the model. And then the same logistic model was developed, but this time with counterfactual sensitive features.

Error for the full model with sensitive features included

```
error = y_test - y_pred_F
```

Testing the full model on counterfactual sensitive data

```
senstest=X_test[["NrOfDependantslessthan3", 'MaritalStatus_1.0', 'MaritalStatus_3.0', 'MaritalStatus_4.0']]
test=X_test[['Age', 'AppliedAmount', 'ExpectedLoss',
                        'LiabilitiesTotal', 'IncomeTotal', 'Interest', 'LoanDuration', 'MonthlyPayment']]

counter_senstest = senstest.replace({0:1, 1:0})
countertest = pd.concat([test, counter_senstest], axis=1)
```

```
y_pred_CF=clf_full_Lin.predict(countertest)
print("For full model:")
print("Mean Squared Error:",mean_squared_error(y_test, y_pred_CF))
print("Root Mean Squared Error:", RMSE(y_pred_CF, y_test))
```

The error for counterfactual linear regression model was computed.

Error for the full model with counter sensitive features included

```
: errorcounter = y_test - y_pred_CF
```

Counterfactual unfairness (CFU) is a computation using difference of both the above-calculated errors.

Counterfactual unfairness (CFU) score for full model

```
: CUF1 = (np.sum(np.square(errorcounter - error))/len(error))
  CUF1
```

Similarly, the error is calculated for an ACF regression-based model with all sensitive feature as is.

Error for the ACF model with residuals of sensitive features

```
acferror = pd.Series(y_test - pred_fair_te, name="Sensitive data")
```

Post which ACF counterfactual residual values (ϵ') are generated. Only the sensitive features are replaced with counterfactual sensitive features (S'). The inversed sensitive features were fit on the already developed models for non-protected independent (X_j) features – the one with original sensitive feature – in prediction mode by calling the model object and executing predict function on it.

Testing the ACF model on counterfactual sensitive data

```
: countersens = sens.replace({0:1, 1:0})
```

The residuals (ϵ') were fit to the already developed model for the response feature (Y) to predict $\left(\widehat{Y'} \right)$ by invoking the model object and using the above residuals. Post which error of the ACF model with counterfactual protected features was calculated.

Error for the ACF model with residuals of sensitive features

```
counter_acferror = pd.Series(y_test - pred_fair_te_acf, name="Counterfactual sensitive data")
```

CUF computation using mean squared difference of errors from ACF model (M_{ACF}) with protected features (S) and counterfactual ACF model (M'_{ACF}) with counterfactual protected features (S').

Counterfactual unfairness (CFU) score for ACF model

```
: CUF2 = (np.sum(np.square(counter_acferror - acferror))/len(acferror))
  CUF2
```

Charts in Fig. 5.11 shocnd the changes we see after implementing counterfactual model.

In case of the first set of plots, logistic regression full model, the density plots show how the error has significantly different distributions when model has factual sensitive features versus when the model had counterfactual sensitive features. But in case of ACF model, there is a good overlap of distribution between errors when model has factual sensitive features versus when the model had counterfactual sensitive features.

For our sample problem, Table 5.20 shows the final results of the CUF calculation for a linear regression model without ACF and an ACF-based model. As you can see, the CUF for the ACF model is significantly lower than that for the linear regression model.

The error of the model between factual and counterfactual increases (which is obvious), but the error between full model and ACF after counterfactual shows ACF-counterfactual performing better than linear model-counterfactual

The result from the CUF technique showed that CUF for a regression model was 5949.80 and CUF for ACF model was 1590.26. The CUF score for regression model is much higher than CUF score of the ACF model, and the distribution of

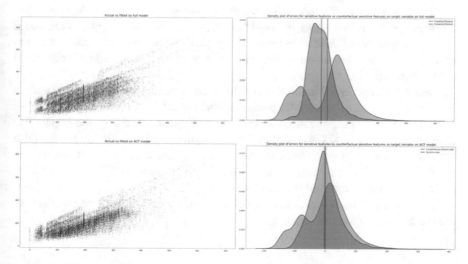

Fig. 5.11 Error vs actual for two models and the impact of counterfactual treatment (red denotes counterfactual sensitive features and blue denotes original)

Table 5.20 Counterfactual unfairness scores for linear vs ACF models

	Linear model		ACF model	
	Factual	Counterfactual	Factual	Counterfactual
MSE	1991.35	7947.53	2442.60	5068.76
RMSE	44.62	89.14	49.42	71.19
CUF	5949.808		1590.26	

predictions for factual and counterfactual in case of ACF is quite similar. This illustrates that the ACF model prediction has very less impact due to the protected features. In other words, the prediction won't change drastically if the fact of sensitive/ protected feature is reversed or is made counterfactual.

Conclusion

The key objective of improving fairness is to remove the discrimination between privileged and unprivileged classes. As we make our predictions fairer, a direct implication is that the false positives and the false negatives both reduce. Consider the example of loan default prediction where a positive outcome is no default and the negative outcome is a default on the loan. The privileged class, as the name suggests has a higher likelihood of a being identified as not going to default on a loan than the unprivileged class. The prediction of whether a person is likely to default on a loan or not is an important input to deciding the outcome of a loan application.

As we make the predictions fairer, this would mean that some of the records from the privileged class who would have had a favourable outcome earlier (no default) would now have an unfavourable outcome (risk of default). This would reduce the number of false positives, loan applicants identified as no risk of defaults, but would have defaulted later.

Similarly, some of the records from the unprivileged class who would have had an unfavourable outcome earlier would have a favourable outcome now. This would reduce the number of false negatives, loan applicants who would've had their loan applications denied because they were considered to have a risk of default but would not have defaulted in reality.

The direct commercial implication of this is reduced number of loans that are not paid back (because of lower false positives) and reduced opportunity loss by better identification of loan applications that will be paid back. Generally, the impact due to reduced false positives is considered higher than the impact due to reduced false negatives as lower false positives. In case of lending use cases, it is not uncommon to see a ratio of 7:3 to be used when calculating the total impact that reduced false positives and false negatives would have.

Defining this ratio is useful when calculating the over business impact of reducing the false positives and negatives.

Bibliography

Calders, T., Kamiran, F. and Pechenizkiy, M. (2009) "Building classifiers with independency constraints," in 2009 IEEE International Conference on Data Mining Workshops. IEEE.

Kamiran, F. and Calders, T. (2012) "Data preprocessing techniques for classification without discrimination," Knowledge and information systems, 33(1), pp. 1–33.

Kilbertus, N. et al. (2019) "The sensitivity of counterfactual fairness to unmeasured confounding," arXiv [cs.LG]. Available at: http://arxiv.org/abs/1907.01040.

Kusner, M. J. et al. (2017) "Counterfactual Fairness," arXiv [stat.ML]. Available at: http://arxiv.org/abs/1703.06856.

Chapter 6
Remove Bias from ML Output

Introduction

In the previous chapter, we saw two techniques that help us address the bias either before the model training by adding weights to the records (reweighting) or by adding a step in the modelling process by calculating the residuals (which requires additional model training). Both of these techniques come in when you still haven't trained your model and allow you to build a model that is fair from grounds up. However, these techniques do not help us if we already have models in production that are doing predictions today that may have bias learnt from the historical data used for training them.

In this chapter, we will discuss how you can address the bias after a model has made its predictions. Since this intervention comes at a much later stage in the ML lifecycle, we need an approach that does not depend on feature engineering or model optimization to be able to reduce the bias. Similarly, the approach we take should be able to work on the predictions regardless of the model used to make the predictions, it should be easy to explain, highly intuitive and should also allow checking for and addressing any residual discrimination in spite of fairness treatments applied in the previous stages of the model development and deployment lifecycle. Finally, we also want to make sure that the approach we apply has little impact on the overall accuracy.

Reject Option Classifier

The approach we will discuss in this chapter is called reject option classifier (ROC). The intuition for the approach comes from the paper "Classification with Reject Option".

© The Author(s), under exclusive license to Springer Nature Switzerland AG 2021
S. Agarwal, S. Mishra, *Responsible AI*,
https://doi.org/10.1007/978-3-030-76860-7_6

ROC applies to two class classification (binary) problems. A binary classification algorithm would yield a result where a classifier $f(X)$ returns an output ($\hat{Y}$) between 0 and 1 (predicted probabilities). Conventionally, the decision boundary for the decision is at $P(Y = 0 | X = x) = 0.5$.

$$f(x) = \begin{cases} 0, & P\left(Y = 0 | X = x\right) \geq P\left(Y = 1 | X = x\right) \\ 1, & \text{otherwise} \end{cases}$$

$$or\ f(x) = \begin{cases} 0, & P\left(Y = 0 | X = x\right) \geq 0.5 \\ 1, & P\left(Y = 0 | X = x\right) < 0.5 \end{cases}$$

However, there is another way to perform the classification using a possible result called reject option (R). The reject option expresses doubt and is to be used for the observations that are hard to classify automatically (when the prediction probability is at the decision boundary or very close to it). Devroye, Gyorfi and Lugosi described this in their paper that classifiers are allowed to report "I don't know", expressing doubt, if the observation is too hard to classify.

Thus, with reject option, the decision is analysed differently. The decisions can be placed into two broad categories: hard and soft. The hard decisions are the ones where the prediction probability is sufficiently far away from the decision boundary for us to make a confident position, e.g. $P(Y = 0 | X = x) = 0.1$ or $P(Y = 0 | X = x) = 0.8$, while the soft decision is the one where the prediction probability is too close to make a confident decision, e.g. $P(Y = 0 | X = x) = 0.45$ or $P(Y = 0 | X = x) = 0.51$.

It is important to note that the decision boundary to define the hard and soft predictions would depend on the problem you are going to solve. It is also an important lever available to you as you utilize the ROC to reduce the bias from the predictions your model has made. At this point, let's add another parameter to the decision boundary that we will call as a critical region. The critical region centres around the decision boundary and covers the range of the soft decision. The distance of the edge of the critical region, or the range of the reject option, from the decision boundary will be denoted by θ. The critical region plays an important role in reducing the discrimination and is another important lever available to you.

Using this very idea (Fig. 6.1), reject option classifier (ROC) is implemented for the labels with probability falling into soft boundary region, which we would also call critical region. A critical region would have all those observations that satisfy the below constraint.

$$\max\left[P\left(Y_{fav} | X\right), 1 - P\left(Y_{fav} | X\right)\right] \leq \text{decision boundary} + \theta$$

where, $0.5 < \text{decision boundary} + \theta \leq 1$

For the decision boundary at $P(Y = 0 | X = x) = 0.5$, let's update the decision logic to include θ

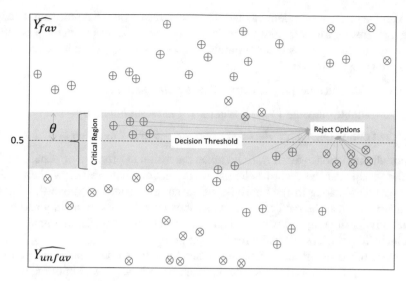

Fig. 6.1 Image depicting parameters of ROC

$$f(x) = \begin{cases} 0, & P\big(Y=0\,|\,X=x\big) > 0.5 + \theta \\ 1, & P\big(Y=0\,|\,X=x\big) < 0.5 - \theta \\ R, & 0.5 - \theta \le P\big(Y=0\,|\,X=x\big) \le 0.5 + \theta \end{cases}$$

Where reject option, denoted by R, lies in the range where we cannot make hard predictions. As covered earlier, θ is the distance of the edge of the critical region from the decision boundary on both sides.

In order to remove discrimination, we would like to minimize the following, within the constraints of the business objectives for the problem.

$$\big|\, P\big(\hat{Y} = \widehat{Y_{fav}}\,|\,S = S_a\big) - P\big(\hat{Y} = \widehat{Y_{fav}}\,|\,S = S_d\big)\,\big|$$

Where $Y \in (Y_{fav}, Y_{unfav})$ are the actual labels with Y_{fav} being favourable class and Y_{unfav} being unfavourable class belonging to dataset X, Y with N data points and $\hat{Y} \in \big(\widehat{Y_{fav}}, \widehat{Y_{unfav}}\big)$ are the predicted labels. $S \in (S_a, S_d)$ is the protected feature with S_a as the privileged/advantaged group, and S_d is the unprivileged/disadvantaged group. When applying ROC, the only data that is accessible to us for the calibration is the predictions of train/test data ($\hat{Y}$), actual values (Y) and protected class (S) – as the model has already been developed or is live in production.

$P(Y|S)$ is the posterior probability of the classifier, which is the probability of Y, given S. Here, if $P(Y_{fav}|X)$ is higher (either close to 1 or 0), then the label assigned would be certain and hard (or vice versa – in case of probability near 0.5). This means the classifier is very certain about the prediction when its falls into the hard boundary region. But the same can't be concluded for the observations falling into soft boundary region.

With the reject option defined, and a quick refresher on what reducing discrimination means, let's look at how can we use it to reduce the bias from the predictions made by any binary classification output. As we make the predictions, our predictions are going to lie in three regions:

1. Favourable outcome $\left(\widehat{Y_{fav}}\right)$, where $\hat{Y} > decision\ boundary + \theta$
2. Unfavourable outcome $\left(\widehat{Y_{unfav}}\right)$, where $\hat{Y} < decision\ boundary - \theta$
3. Reject option (R), where $decision\ boundary - \theta \le \hat{Y} \le decision\ boundary + \theta$

For all predictions within R, ROC takes the action to promote or penalize based on the group that the record belongs to. Prediction that are below the decision boundary and belong to the unprivileged group are promoted above the decision boundary, whereas the predictions that are above the decision boundary and belong to the privileged group are penalized and moved below the decision boundary. This results in "flipping" the decision for the two groups.

Figure 6.2 demonstrates ROC in action. As you can see, the privileged class within the ROC is moved below the decision boundary (as we apply the penalty), whereas the unprivileged class is promoted above the decision boundary (0.5 in this case).

Any prediction outside the critical region won't get impacted. Thus, only selected soft probabilities would get treated, and all other probabilities would stay as-is.

Note: When Can You Use ROC?
Reject option classifier is a highly intuitive method to reduce the bias in predictions. Since it is applied after the predictions are made, the questions that can be asked is: Why do we need to apply it? Or how do we identify the right time to apply ROC?

To answer these, let's begin by summarizing the key strengths of ROC:

- It works with any probabilistic classifier with binary classification or any continuous output as long as you can put a decision boundary to do binary classification.
- It requires neither modification of learning algorithm nor pre-processing.
- It gives better control and explainability for bias reduction.

For any models that are already in the production environment, ROC can come in and help reduce any bias in the predictions with a minimal effort as compared to redesigning your model from the ground up. Another application of ROC is when you are using a black box model to make the predictions. It cannot help explaining the predictions of a black box model, but the actions taken by the ROC technique remain explainable.

We have limited our treatment of ROC to problems of binary classification – where the output is one of the two possible options. As long as you can make your prediction into a binary output and have a decision boundary within the range of predictions (probabilistic or continuous output), you can apply ROC.

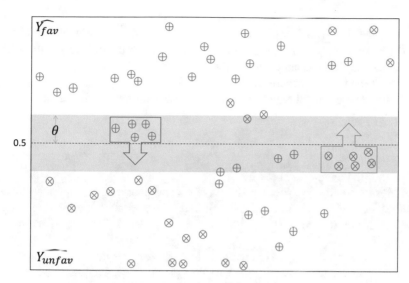

Fig. 6.2 Image depicting movements of observation during ROC

The ROC algorithm would therefore, for all probabilities outside critical region defined by theta, keep the label or the probabilities unchanged, but for probability falling into critical region, the label or the probabilities assignment would be affected as per the below rule:

$$\hat{Y} = \begin{cases} \widehat{Y_{fav}}, & \text{if } \widehat{Y}_i \in S_d \\ \widehat{Y_{unfav}}, & \text{if } \widehat{Y}_i \in S_a \end{cases}$$

Thus, if these reject option instances (falling into our critical region) are from disadvantageous group, then they are labelled with favourable outcome (Y_{fav}), and if the observations are from advantageous group, then they are labelled with unfavourable outcome (Y_{unfav}). However, here the labels won't be changed from 0 to 1 or vice versa, but the probabilities would be changed so that its falls into either Y_{fav} or Y_{unfav} class as required. The instances outside the critical region (θ) are classified according to the standard decision boundary definition.

The implementation of ROC may seem quite trivial, but as they say that devil is in the details. The ROC implementation can go haywire if it's not carefully written, especially the logic and condition of ROC treatment. Below are the high-level steps for understanding the ROC treatment before we look at the code for the implementation of ROC and the fairness metrics for our example:

1. Get actual labels, predicted labels and predicted probabilities.
2. Find out the relevant protected features that need treatment.
3. Create composite protected features, if necessary.
4. Create copy of predictive probabilities.
5. Filter the data falling in critical region.
6. Filter the data for each of four protected and predicted class combination ($Y_{fav} \wedge S_a, \widehat{Y_{fav}} \wedge S_d, \widehat{Y_{unfav}} \wedge S_a, \widehat{Y_{unfav}} \wedge S_d$) in the critical region.
7. Swap the probabilities.

(a)   For $\hat{Y} = \widehat{Y_{unfav}} \wedge S_d$, swap it to Y_{fav} by changing the probability to $(1-\hat{Y})$.

(b)   For $\hat{Y} = \widehat{Y_{fav}} \wedge S_a$, swap it to Y_{unfav} by changing the probability to $(1-\hat{Y})$.

8. Check the model accuracy after ROC treatment.
9. Check the fairness accuracy.
10. Optimize the critical region to reduce accuracy-fairness trade-off.

Note: The Parameters of ROC

There are three parameters that govern the output from the application of ROC. Before we will see them in action across different examples, let's try and understand them and their intuitions in a little detail. The three parameters are as follows.

Decision boundary is the separating line between the favourable and unfavourable outcomes. The outcomes that lie exactly on the decision boundary can belong to one of the outcomes – as defined by you.

The examples in the chapter use the value of 0.5 for the decision boundary, but that is illustrative, and you should decide the value based on the distribution of your data.

Theta (θ) decides the critical region – the belt above and below the decision boundary. All outcomes that satisfy either of these two conditions $(\widehat{Y_{unfav}} \wedge S_d, \widehat{Y_{fav}} \wedge S_a)$ and fall within the critical region are impacted by the ROC.

For all the examples covered in the chapter, the critical region extends by the same amount in both directions of the decision boundary – however, if you feel necessary, you may define two versions of θ. The distance by which the critical region extends into the unfavourable outcome territory can defined as θ_d – because this will impact the unprivileged class S_d identified for promotion, and the distance by which it extends into the favourable outcome as θ_a, which will identify the privileged class S_a for penalty – if applied.

ROC performs action to promote $\widehat{Y_{unfav}} \wedge S_d$ or penalize $\widehat{Y_{fav}} \wedge S_a$. We expect nearly all flavours of ROC to include promoting the unfavourable outcome for the unprivileged group to reduce the bias, but the decision to penalize will depend on the specific problem you are trying to solve and the associated fairness objectives.

Shown below is an example where the first image shows the outcomes on the two sides of the decision boundary (all unfavourable outcomes are coloured blue with hollow squares representing S_a and solid squares representing S_d). In this example, decision boundary is at 0.5, θ is 0.1, and the action being taken is to promote only, with no penalty. You can see the movement of the unprivileged class across the decision boundary as ROC gets applied.

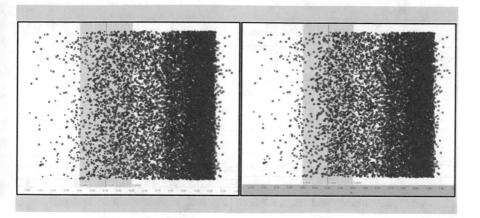

Let's continue with our example of loan data. In the illustration, the subset of data was used with (X_j) being 12 independent features including 4 protected feature (S) viz. Married, Single, Divorced and No. of Dependants more than three. The predictions from logistic regression model (M_{LR}) were used for ROC optimization. Here we would only be using the model output and the original data, and there would be no requirement of model or model parameters.

In first few initial steps, we declare the decision boundary, θ (theta), privileged group values (S_a), favourable group values (Y_{fav}) and protected class (S).

```
D_Boundary = 0.5
theta = 0.1
penalise = True
pval = 1
upval = int(not pval) #Unprivileged
```

We then create a combined data frame with all the data we will be using for ROC algorithm. Here, we also create a copy of original predicted probabilities which we will be using in our ROC calibration.

```
roc=pd.DataFrame()
roc_df = pd.DataFrame(pred_prob, columns=['pred_prob'])
roc_df['y_pred_original'] = y_pred_prob
roc_df['y_pred'] = y_pred
roc_df['y_test'] = y_test.reset_index(drop=True)
roc_df['S'] = X_test[choice].reset_index(drop=True)
roc_df['id']=roc_df.index
roc_df.head()
```

	pred_prob	y_pred_original	y_pred	y_test	S	id
0	0.851482	0.851482	0.0	1.0	1.0	0
1	0.828825	0.828825	0.0	0.0	1.0	1
2	0.831863	0.831863	0.0	0.0	1.0	2
3	0.665093	0.665093	0.0	0.0	0.0	3
4	0.767699	0.767699	0.0	1.0	1.0	4

Now we need to create a subset of only those records that fall into our critical region. For the problem at hand, we want to ensure that the critical region is on both side of decision boundary. As discussed earlier, if you choose to apply ROC on a problem, this decision should lie largely with business analyst and the product owner.

```
if penalise==True:
    RO_df = roc_df[(np.abs(pred_prob - 0.5) < theta)]
else:
    RO_df = roc_df[(np.abs(pred_prob - 0.5) < theta) & (y_pred==fav)]
```

```
RO_df.head()
```

	pred_prob	y_pred_original	y_pred	y_test	S	id
6	0.499860	0.499860	1.0	0.0	0.0	6
7	0.507807	0.507807	0.0	0.0	0.0	7
10	0.476438	0.476438	1.0	1.0	0.0	10
11	0.596510	0.596510	0.0	1.0	0.0	11
15	0.595882	0.595882	0.0	0.0	0.0	15

The following steps are where the ROC gets implemented. We would be, for ease, creating subset of data for all combinations viz. favourable and privileged $\left(\widehat{Y_{fav}} \wedge S_a \right)$, favourable and unprivileged $\left(\widehat{Y_{fav}} \wedge S_d \right)$, unfavourable and privileged $\left(\widehat{Y_{unfav}} \wedge S_a \right)$ and unfavourable and unprivileged $\left(\widehat{Y_{unfav}} \wedge S_d \right)$ and perform fairness treatment on each subset of the data separately.

The predicted probabilities (copy of the original predicted probabilities) of required set of data would be swapped by subtracting the original predicted probabilities from 1. As discussed earlier, the unprivileged group (S_d) of unfavourable class (Y_{unfav}) would be sent to the favourable side (Y_{fav}) of decision boundary. Likewise, the privileged group (S_a) of favourable class (Y_{fav}) would be sent to the unfavourable (Y_{unfav}) side of decision boundary since we opted to also penalize the observations. Then all that remains is combining all the subsets of data and replacing the new values within the main dataset for further analysis.

```
pval_fav = RO_df[(RO_df['S']==pval) & (RO_df['y_pred']==fav)]
pval_ufav = RO_df[(RO_df['S']==pval) & (RO_df['y_pred']==unfav)]
upval_fav = RO_df[(RO_df['S']==upval) & (RO_df['y_pred']==fav)]
upval_ufav = RO_df[(RO_df['S']==upval) & (RO_df['y_pred']==unfav)]
```

```
pval_fav['pred_prob'] = 1- pval_fav['pred_prob']
upval_ufav['pred_prob'] = 1-upval_ufav['pred_prob']
```

```
RO_changed = pd.concat([pval_fav, pval_ufav, upval_fav, upval_ufav])
```

```
RO_changed.head()
```

	pred_prob	y_pred_original	y_pred	y_test	S	id
53	0.480051	0.480051	1.0	0.0	1.0	53
5906	0.460625	0.460625	1.0	0.0	1.0	5906
7	0.507807	0.507807	0.0	0.0	0.0	7
11	0.596510	0.596510	0.0	1.0	0.0	11
15	0.595882	0.595882	0.0	0.0	0.0	15

```
df = roc_df
```

```
df.loc[list(RO_changed.index), 'pred_prob']=RO_changed['pred_prob']
```

With the ROC implemented, let's look at the impact of the ROC techniques on the discrimination and the fairness metrics due to the protected features

Protected Feature: Single

AUC scores for the protected class "Single" is very close to the AUC score (Table 6.1) of original models; this means that there is a very slight decay in model accuracy. When we calculate the fairness metrics, we see a significant improvement in most of the fairness metrics. In this case, the theta value was 0.1 and there was no penalization.

Also, for this protected feature, the difference in other accuracy metrics (AUC, precision and TPR) before and after applying ROC (Fig. 6.3 and Table 6.1) has narrowed largely showing a positive impact of the algorithm.

Table 6.1 AUC values before and after ROC treatment

AUC ROC	AUC Original	Ratio
0.6849	0.6865	0.9976

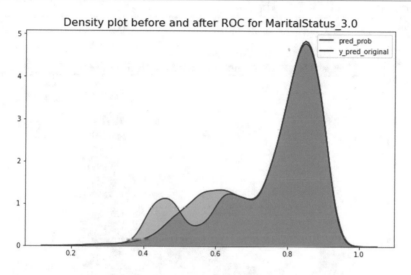

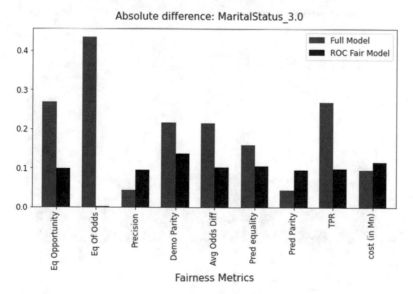

Fig. 6.3 Density plot and difference in fairness metrics for protected feature Single

Protected Feature: Married

For the protected class "Married", the AUC scores (Table 6.2) remain almost identical before and after the ROC treatment. However, the fairness metrics show an improvement across the board. The ROC treatment in this case is different as we have used $\theta = 0.01$ (critical region between $0.49 \leq \hat{Y} \leq 0.51$) and are penalizing and promoting. The results for the parameters we have chosen in this example show an excellent improvement (Fig. 6.4) in fairness for equal opportunity and TPR. The cost unlike the last example remains same before and after ROC treatment.

Table 6.2 AUC values before and after ROC treatment

AUC ROC	AUC original	Ratio
0.6865	0.6865	0.9999

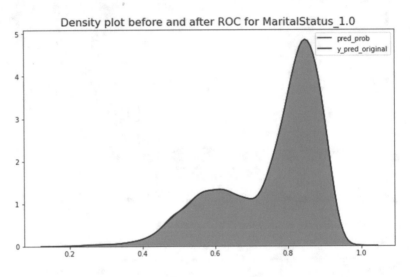

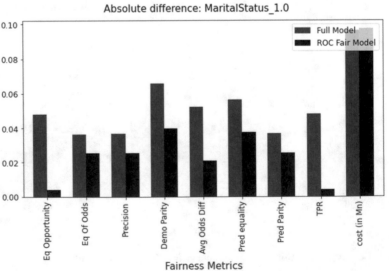

Fig. 6.4 Density plot and Difference in fairness metrics for protected feature Married

Protected Feature: Divorced

For the protected class "Divorced", the AUC score (Table 6.3) before and after ROC treatment shows a negligible decay. However, the impact of ROC treatment is quite significant for all fairness definitions. Even the secondary model metrics show (Fig. 6.5) a very good improvement with respect to gap between the two groups of the protected class. In case of Divorced, the parameters for ROC treatment had a theta value of 0.1 with penalization for advantageous group.

Table 6.3 AUC values before and after ROC treatment

AUC ROC	AUC Original	Ratio
0.6865	0.6865	0.9985

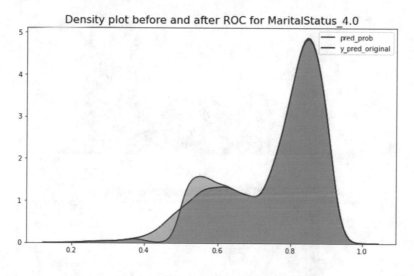

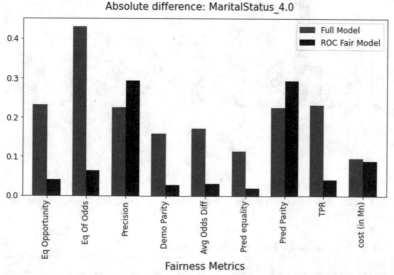

Fig. 6.5 Density plot and Difference in fairness metrics for protected feature Divorced

Protected Feature: Number of Dependants Less than Three

Again, here we use a different range for deciding the critical region as we use $\theta = 0.1$. The plots (Fig. 6.6) are for the protected feature representing if "No. of dependants less than three". Like the previous examples, the AUC score (Table 6.4) is very close to the original AUC score showing no/less decay in model. With the

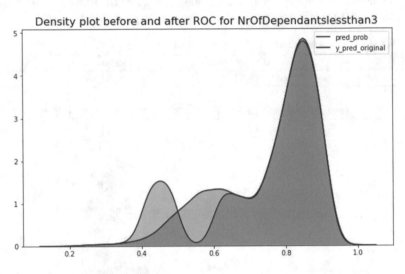

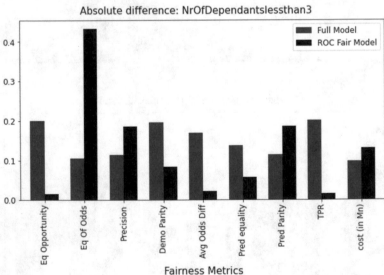

Fig. 6.6 Density plot and Difference in fairness metrics for protected feature Number of dependants less than three

Table 6.4 AUC values before and after ROC treatment

AUC ROC	AUC Original	Ratio
0.6828	0.6865	0.9946

parameters ($\theta = 0.1$ and penalization) used, we are not able to optimize the ROC at its fullest, and thus we see only two major fairness metrics (equal opportunity and demographic parity) had positive impact, but all other accuracy model metrics saw a significant improvement. As said earlier, sometimes it can be really difficult to optimize all fairness metrics, and the prioritization is largely dependent on the business requirements and objectives for the problem that you are trying to solve.

Optimizing the ROC

As we have seen above, different protected features behave differently, and thus it gets important to optimize the parameters to ensure minimum decay in accuracy and maximum increase in fairness across all or prioritized protected features. The ROC technique can be optimized using following steps:

1. Optimizing the decision boundary – chose decision boundary of the classifier other than 0.5 default value.
2. Optimize critical region (theta value).

    (a) Choose θ value separately promoting the unprivileged class S_d (denoted by θ_d) and for penalizing the privileged class S_a (denoted by θ_a).

For instance, if we aim to go with step 2 as described above (while keeping the same value of θ for both sides of the decision boundary), we can plot the theta vs the fairness metric to choose optimal theta. In the plot (Fig. 6.7), we see that we predicted all possible fairness metrics for different values of theta (critical region) ranging from 0 to 1.

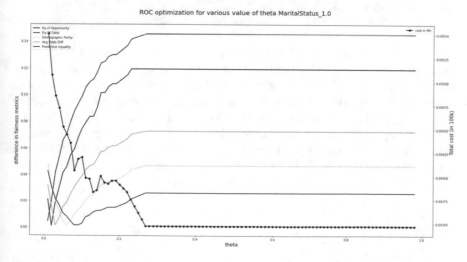

Fig. 6.7 Plot showing values of different fairness metric across range of theta

It is evident that for minimum cost (just to reiterate, cost here is deduced as $FP \times 700 + FN \times 300$), the theta should be near 0.25 (and not 0.10 ideally), but to minimize discrimination with respect to the demographic parity, theta should be around 0.05 (and not 0.10). But then if the choice of fairness metrics is different, then the choice of critical region needs to be altered likewise.

Handling Multiple Features in ROC

So far, we have seen how to bring down discrimination due to a single protected feature. If we are lucky, one single ROC can treat discrimination by all protected features in one single shot. Or else we can opt for the following two options:

1. Sequentially treat multiple features on the ROC treatment of previous protected feature.
2. Go for composite feature.

We have already introduced the concept of composite feature in the previous chapter. Here again the idea of composite sensitive features would come in place, and as described before, this can be defined using logical operator. In this case, the composite feature was defined using logical operator on "No. of Dependants less than three", "Divorced" and "Work Experience less than 10 years" protected features. This composite feature was a combination of No. of Dependants less than three, Divorced and Work Experience less than 10 years with disadvantageous group being defined with OR operator.

```
data['dep_MS4_wex10']  =  data[['NrOfDependantslessthan3','MaritalSta
tus_4.0', 'WrExLess10']].min(axis = 1)
```

Here the observations falling between the probability range of 0.40 to 0.60 were included in the critical region (θ being 0.1) with advantageous observation getting penalized, and we see some interesting results. There is relatively no decay in AUC scores (Table 6.5), and all-important fairness metrics (Fig. 6.8) saw a very significant improvement; even the secondary model metrics saw decrease in discrimination between two groups of these newly defined composite feature. It would be interesting to further optimize the parameters for further improvement. But then it is important to set the threshold for discrimination initially by the business analysts and product owner or else this can be a never-ending exercise.

Table 6.5 AUC values before and after ROC treatment

AUC ROC	AUC original	Ratio
0.6855	0.6865	0.9985

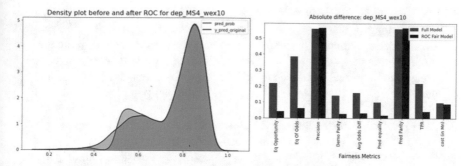

Fig. 6.8 Density plot and difference in fairness metrics for protected composite feature

Conclusion

Before we close the chapter, let's just discuss ROC implementation in production (Fig. 6.9). In most of the cases in production, we won't have the actual labels and would only have the predictions and protected class. In such cases, the historical data should be used to learn the ROC model and its optimal parameters. Post-multiple iterations, the θ (critical region range) and decision threshold should be calculated which then can be applied to production data. However, it would be really important to keep a track of fairness metric regularly on the production data, and any deviation would mean recalibrating the theta and decision boundary.

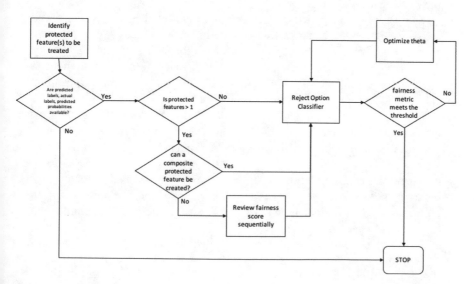

Fig. 6.9 Flowchart for applying reject option classifier to remove bias from the model output

Note: Extending ROC

As we close this chapter, we want to introduce two new thoughts. At the time of completing the manuscript, we have not had the time to do a deep dive and research more in these, but we would like to propose them as we believe that there is a merit in these ideas and can be useful in applying ROC to different problems.

Promoting and Penalizing: When promoting or penalizing a record, the approach we have taken so far is to update the predicted probability to $\left(1-\hat{Y}\right)$. This moves the record to the "other side" of the decision boundary and "flips" the outcome from favourable to unfavourable (when penalizing) and from unfavourable to favourable (when promoting). Consider the scenario where you have multiple $\overbrace{Y_{unfav}} \wedge S_d$ records within the critical range that you are going to promote; the approach of changing the probability to $\left(1-\hat{Y}\right)$ does not keep the relative positions of these records intact. For example, if you have two records with probabilities 0.41 and 0.49, they become 0.59 and 0.51. While the outcome for both of them is now favourable and the same, the one that had a higher score before ROC was applied now has lower score. This is an unintended consequence of the process we have applied.

We know that θ is the distance between the decision boundary and the edge of the critical range. Another approach we can take to flip the records to the other side of the decision boundary while keeping their relative positions intact is by adding or subtracting θ from the predicted probability depending on whether we are promoting or penalizing the record. Thus, when promoting a record, we will change the probability to $(\hat{Y}-\theta)$, whereas when penalizing it will change to $\left(\hat{Y}+\theta\right)$.

Since the fairness metrics depend on the total number of records moved across the decision boundary, this change in the approach will have hardly any, if at all, impact on them. When we tried this approach for the features demonstrated in the chapter, there was no difference in the fairness metrics. There is some difference in the AUC and the AUC ratio as a result. Depending on the feature and the underlying distribution, the AUC ratio might improve to a higher coverage (it changed from 0.99462 to 0.99773 for "No. of Dependants less than 3") or go down a bit. You may want to consider that as you decide the mechanism to promote and penalize the records for your problem.

Multi-class Classification: In this chapter, we have used ROC only for binary classifications. We believe that ROC's application in multi-class problems as a single technique to handle all classes together is a challenge worth exploring and needs more research.

If you have a multi-class problem with distinct classes, it is worth exploring if you can break it down into multiple binary classifications and treat the discrimination within each using ROC. When trying this approach, it will be important to spend time defining the privileged and unprivileged classes and the favourable and unfavourable outcomes for each of the binary classifications.

Bibliography

Bantilan, N. (2021). cosmicBboy/themis-ml. [online] GitHub. Available at: https://github.com/cosmicBboy/themis-ml [Accessed 16 Apr. 2021].

Herbei, R. and Wegkamp, M.H. (2006). Classification with reject option. *Canadian Journal of Statistics*, 34(4), pp.709–721.

Kamiran, F., Karim, A. and Zhang, X., 2012. Decision Theory for Discrimination-Aware Classification. 2012 IEEE 12th International Conference on Data Mining,.

Chapter 7
Accountability in AI

Introduction

Imagine spending long nights through the model development lifecycle, ensuring that your data represents the real world, is free of bias, the model explainable and the accuracy high. You push the model to the production, and everything looks great. The performance in the production environment is as expected, and your data scientists can now work on the next problem. However, as the time goes by, you notice that the model performance has started to dip – not just leading to lower accuracy but also lower performance on the fairness metrics. How is this possible and what did you not account for?

The issue we've just described is not about errors in model development but instead about what comes next. For majority of the problems, the data distribution of the live data shifts over a period of time. This is known as data drift (or model drift), and if left unattended, it leads to performance degradation of the model. From a responsible AI point of view, this also means that in addition to lower overall accuracy, the model will also perform differently on the key fairness metrics that you have identified. Another common reason for the poor performance on the production environment is production skew – which is the difference in the model performance between training and production environments. Production skew can happen because of errors in training, bugs in the production environment or because the training data and the live data do not follow the same distribution. This, just like data drift, leads to lower overall performance – from accuracy to fairness metrics.

This brings us to the next stage of the model's lifecycle – model monitoring, since it is not responsible AI if the model is penalizing/rewarding incorrectly or making wrong and absurd decisions for the end-user post-deployment. Model monitoring is a group of techniques and activities we need to perform to identify any drift that can impact the performance of the model and to prevent it. To some extent, it is the most important part of data science which actually starts after model development.

To be able to monitor a model's performance, the first and the foremost challenge is quantifying the model degradation. Identifying the parameters to track the model performance and defining the thresholds that if breached should raise an alert are fundamental components of model monitoring. This is important as it allows us to avoid spending more effort than required on retraining the models and yet ensure that we are not performing below the standards we have set for ourselves.

In this chapter, we will cover the following topics:

- Ensuring a model to remain true to its purpose
- Detecting drifts during deployment/live
- Detecting any significant change due to:

  • Drift in X (features) or Y (prediction) called data drift
    ◦ Drift in Y is also called prior probability drift
  • Drift in $P(X|Y)$ called concept drift

- Ensuring fairness and XAI to be always valid
- Reducing cost by alarming on wrong decisions (predictions)

You may choose to leverage most of these techniques as part of model monitoring, but at the minimum, it is imperative to have a process for detecting, alerting and addressing any kind of drift that may occur post go live. Eventually a model that exhibits any kind of degradation will need to be examined further either for recalibration, retraining or, in the worst case, replacement.

Let's look at the different measures and approaches. Some of them are associated with code in the chapter; these are the ones where we have combined techniques or are not available readily as part of the standard packages. For other measures or approaches that are notable in the value, they provide and are included in standard packages; the associated code is included in the chapter's notebook in the associated repository.

Regulatory Guidelines

With the increasing adoption of the machine learning, the awareness of the impact it can make and the need to better understand and monitor how it works is increasing too. The regulators, especially the financial regulators, have started taking note of this, and we can now see that the regulators have started creating guidelines for model monitoring. BASEL has strict guidelines for model monitoring especially for credit risk models. Even Board of Governors of the Federal Reserve System has guidelines on model monitoring. The Fed SR 11–7 recommends that:

- Model is being used and is performing as intended.
- Monitoring is essential to evaluate whether changes … necessitate adjustment, redevelopment or replacement of the model and to verify that any extension of the model beyond its original scope is valid.

- Monitoring should continue periodically over time, with a frequency appropriate to the nature of the model, the availability of new data or modelling approaches and the magnitude of the risk involved.
- Sensitivity analysis and other checks for robustness and stability should likewise be repeated periodically.
- If models only work well for certain ranges of input values, market conditions or other factors, they should be monitored to identify situations where these constraints are approached or exceeded.
- Discrepancies between the model output and benchmarks should trigger investigation into the sources and degree of the differences and examination of whether they are within an expected or appropriate range given the nature of the comparison. The results of that analysis may suggest revisions to the model.

Data Drift

Let's say you train a model to predict the price of a cup of coffee given the neighbourhood – you train the model using historical data over a period of few years and want to see how your model performs for the years after that. You've even taken care of using inflation to price adjust to give more accurate price estimate. Your model seemed to work fine for the initial years, but as you move forward in the time, you find that for some areas it tends to underestimate the price of a cup of coffee.

As you analyse, you realize that new offices opened up close to most of these neighbourhoods and that gave the coffee shops a steady supply of customers during the day. With a mostly captive audience, the coffee shops could afford to charge a little more for a cup leading to underestimation by your model.

What you experienced is data drift, where the underlying distribution of the live data changes from the distribution of the data used for training the model causing poorer model performance. Just as the second law of thermodynamics states that the entropy of a system left to spontaneous evolution always increases, all machine learning models decay over the time as the underlying distribution changes. Given that economic conditions and demographics are becoming extremely volatile and dynamic, it would be naïve to assume that a model developed a while ago would hold its fort in the current situation as well.

Detecting drift requires continuous monitoring to gauge if the drift is one-time, sporadic or regular one. The data drift, however, relatively speaking is the easier one to detect. Since data drift is caused by the change in the underlying distribution, we just need to monitor the distribution of the live input data and compare it with the training data to identify any drift. This does not require us to know the ground truth (actual outcome), and hence theoretically we can detect the data drift as it happens. Let's have a look at some measures/techniques that can help us detect the data drift.

Covariate Drift

Covariate drift is the change in the distribution of the input feature set (or covariates – input variables that have a relationship with the dependent variable, the prediction). This is one of the most common causes of model drifts, and the distribution drift can happen due to many reasons. The drifts can happen slowly over a period of time, for example, the shopping patterns for technology products, or relatively quickly when new products or services are introduced, like changes in vacation spending patterns due to Airbnb and similar businesses. A lot of time drift in data distribution occur because of latent changes, macro-economic changes or demographic changes – which can be very difficult to detect on time.

As the covariate drift happens, the distribution of the live data shifts from that of the data used for the training and testing. The distance between the non-intersection of the two distributions is a very good measure of the drift.

$$d(P, Q) = 1 - \sum_i \min(P_i, Q_i)$$

Where P and Q are training and live distributions, respectively. The larger the distance, the bigger is the drift. To calculate this, each independent variable is binned to form i bins (commonly 20 equal bins) from both the actual distribution and live distribution taken together. Then, the shift in the variable contribution to each bin is calculated.

The function below picks the common variables between the two datasets and calls the calculate_distance function to get the values for each variable.

```
def calculate_covariate_drift(data_train, data_live, bins=20):
    # variables present in both datasets
    joint_var = data_train[data_train.columns.intersection(data_live.columns)].columns
    # distances between variables
    distances = int(len(joint_var))

    distances = {}

    for x in joint_var.values:
        distances[x] = dict(value=calculate_distance(data_train[x], data_live[x], bins))

    return distances
```

We begin by sorting the distributions by data rank and then creating bins of equal sizes. For a given feature, the percentage of observations falling into each bin is computed separately for the training and live data. The distance calculation after that is straightforward – the sum of the minimum percentage across each bin is calculated and then subtracted from 1. Figure 7.1 shows the output from the function calculate_distance.

```
def calculate_distance(data_train, data_live, bins = 20):
    col=data_train
    if isinstance(col,int):
        com_list = list(data_train) + list(data_live)
        com_list = pd.DataFrame(com_list)
        rank = com_list.rank()
        after_cuts = pd.DataFrame(data_train, data_live)
    else:
        com_list = list(data_train) + list(data_live)
        com_list = pd.DataFrame(com_list)
        rank = com_list.rank()
        after_cuts = pd.cut(np.array(rank).ravel(), bins)

    l1 = list(np.repeat(1, len(data_train)))
    l2 = list(np.repeat(2, len(data_live)))
    master_list = l1 + l2
    x = pd.crosstab(after_cuts, pd.Series(master_list))
    matrix = x / x.sum()

    value = 1 - np.sum(matrix.apply(np.min, axis=1))
    return value
```

col_0		1	2	col_0		1	2
row_0				row_0			
(-45.674, 3048.7]	1942	825		(-45.674, 3048.7]	0.045243	0.044844	
(9116.1, 12149.8]	9693	4256		(9116.1, 12149.8]	0.225818	0.231342	
(15183.5, 18217.2]	1195	494		(15183.5, 18217.2]	0.027840	0.026852	
(18217.2, 21250.9]	2123	912		(18217.2, 21250.9]	0.049460	0.049573	
(21250.9, 24284.6]	1321	561		(21250.9, 24284.6]	0.030775	0.030494	
(24284.6, 27318.3]	2749	1163		(24284.6, 27318.3]	0.064043	0.063217	
(27318.3, 30352.0]	2104	850		(27318.3, 30352.0]	0.049017	0.046203	
(30352.0, 33385.7]	2327	1003		(30352.0, 33385.7]	0.054212	0.054520	
(33385.7, 36419.4]	1853	826		(33385.7, 36419.4]	0.043169	0.044899	
(36419.4, 39453.1]	2257	1000		(36419.4, 39453.1]	0.052581	0.054357	
(39453.1, 42486.8]	2316	904		(39453.1, 42486.8]	0.053956	0.049138	
(42486.8, 45520.5]	1047	479		(42486.8, 45520.5]	0.024392	0.026037	
(45520.5, 48554.2]	3107	1741		(45520.5, 48554.2]	0.072384	0.067457	
(48554.2, 51587.9]	2818	1253		(48554.2, 51587.9]	0.065651	0.068109	
(51587.9, 54621.6]	1264	552		(51587.9, 54621.6]	0.029447	0.030005	
(54621.6, 57655.3]	2292	953		(54621.6, 57655.3]	0.053397	0.051802	
(57655.3, 60689.0]	2516	1125		(57655.3, 60689.0]	0.058615	0.061151	

Fig. 7.1 (L to R) frequency of data in each bin for train and live sets, proportion of data in each bin for train and live sets

Using a predefined threshold, alerts are generated by passing the distance calculated above for each batch of new or live data. The thresholds used below are representative, and you may want to adjust them based on your data.

```
def cal_threshold(d):
    for k, v in d.items():
        if v["value"] > 0.40:

            d[k]['Thresh'] = "High Drift"
        elif v["value"] > 0.30:

            d[k]['Thresh'] = "Medium Drift"
        elif v["value"] > 0.20:

            d[k]['Thresh'] = "Low Drift"
        else:
            d[k]['Thresh'] = "No Drift"
    return d
```

Continuing with the same dataset, for the feature "Applied Amount", the covariate drift over 30 days (Fig. 7.2) using the above concept shows a need for investigation as the drift is close to the High Drift level on a large number of days.

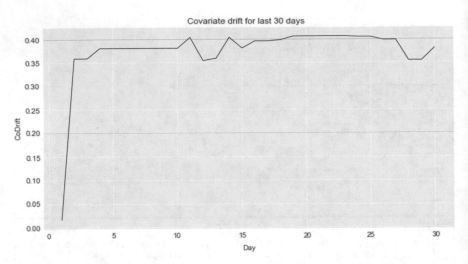

Fig. 7.2 Covariate drift for Applied Amount for last 30 days along with Red, Amber, Green (RAG) status as declared in code above

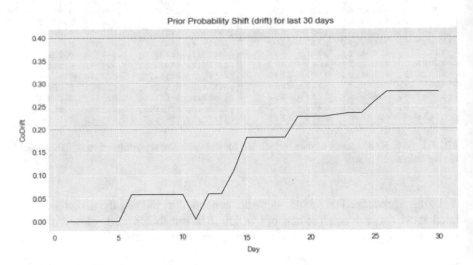

Fig. 7.3 Prior probability shift for target feature using PSI metric

Extending this further, the same concept can be even replicated for the outcome – the predicted probabilities. The same techniques applied on the predicted probabilities is called prior probability shift (Fig. 7.3). This refers to the change in the distribution of the target variable in the training data and the live data. The target variable is binned to form i bins from both the distribution. Then, the shift in the variable contribution to each bin is calculated.

The only shortcoming of this method is its heavy reliability on the way binning is done. Thus, it makes it very dependent on size and number of bins. Moreover, given that it gives equal weightage to all deviations, a bin with small deviation in numbers of observations – but with significant deviation in intra-bin distribution – won't get highlighted. Furthermore, if the bin size in the live data gets small (say gets reduced significantly), the drift would go unnoticed as it would get out-weighted by other heavy frequency bins.

Jensen-Shannon Distance

The covariate drift is a slightly naïve approach considering it computes drift based on a simple distance measurement. Let's look at another approach to measure divergence between the training and live distributions – Jensen-Shannon Distance. In Chap. 3, Fairness and Proxy Features, we talked about mutual information, which measures the amount of information one can obtain from one random variable given another. Jensen-Shannon distance is the mutual information for a variable associated to a mixture distribution between P and Q, where P and Q are as described earlier.

Jensen-Shannon distance is based on another measure, the Kullback-Leibler divergence, which is often used for detecting the randomness in continuous time series and computing the relative entropy in information systems. We are discussing JS distance here because it is symmetric and has a finite value. The JS distance between two probability vectors is defined as

$$JSD(P \| Q) = \sqrt{\frac{D(P \| m) + D(Q \| m)}{2}}$$

where $m = \dfrac{P+Q}{2}$

The divergence (D) in JSD is calculated using K-L divergence formula between two distribution given by

$$D_{KL}(P \| Q) = \sum_i P_i \log \frac{P_i}{Q_i}$$

If you read more about Jensen-Shannon distance, you will come across JS divergence. The distance is related to the divergence and is a square root of the divergence. Therefore, JS divergence is defined as

$$\text{JSDivergence}(P \| Q) = \frac{D(P \| m) + D(Q \| m)}{2}$$

JSD's scores lies between 0 (no change in distribution) and 1 (significant change in distribution). The JSD can also be computed for multidimensional distributions, thus making it fit for multi-class model monitoring use cases too. The chart in Fig. 7.4 shows how the JSD value increases over a period of time. Based on your

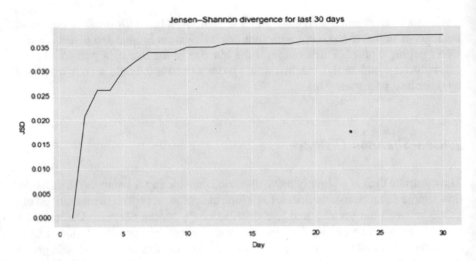

Fig. 7.4 JSD for Applied Amount comparing training distribution and live data distribution for last 30 days

data and the problem, you can define the thresholds to identify when the drift is large enough to warrant more analysis.

Wasserstein Distance

Another measure of distribution drift is Wasserstein distance. It is defined as the minimum amount of work required to move one distribution over to the other distribution. Unlike measures for distributions distances, it is ideal for measuring shift in all types of data. To understand the intuition behind Wasserstein distance, consider the three sample distributions shown in Fig. 7.5:

The first image shows an example of the original distribution followed by two examples of the drift. If we apply either of the previous two measures of distance (covariate drift or Jensen-Shannon), the distance between the original and case 1 will be the same as original and case 2. However, intuitively, you can see that the case 2 is further apart from the original distribution than the case 1. The amount of work needed to move the original distribution over to case 2 is more than the amount of needed to move it to case 1's distribution.

This effort is what Wasserstein distance captures. It is much easier to visualize (Fig. 7.5) by looking at the plot of the distribution for the example, but the real-world data for almost any problem is not going to be as easy to visualize manually.

In our context, the Wasserstein metric measures the probability and the distance between various outcomes, in addition to the similarity of distribution rather than

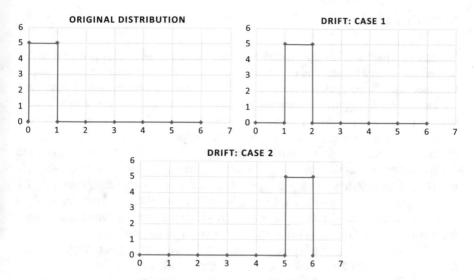

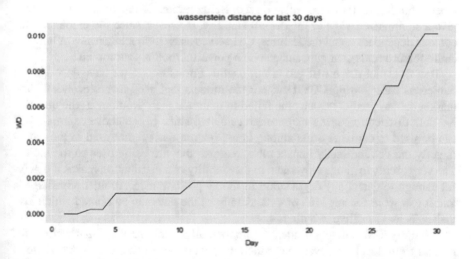

Fig. 7.5 On the left is the original distribution followed by two examples of drift

Fig. 7.6 EMD for Applied Amount comparing training distribution and live data distribution for last 30 days

exactly matching the distributions. We can use it to measure the similarity for both numerical and categorical features as it is not bin dependent but depends on overall distribution and computes the score based on difference in the shape and distance of the two distributions.

Figure 7.6 shows the increasing Wasserstein distance between training and live data over a period of 30 days for a sample dataset.

Stability Index

Validity of a model is highly dependent on the similarity between the data distribution on which it is trained and the live data on which it makes its predictions. As the live data distribution changes, the validity of the model can come under the scanner. As the underlying population changes, there comes a point where it is safe to assume that the model is no longer fit for the purpose and its predictions no more reliable.

Stability index is a measure of change of the population of a given feature, be it for the predictor or for the response variable – it computes the percentage change in the population across time period. There are two measures that we will talk about in this section, population stability index (PSI) and characteristic stability index (CSI). An advantage of the stability index is that we don't need the actual outcomes of the live data (or the ground truth) to be able to conclude if the model which was trained on historical demographic is still true for the current demographic.

PSI measures the percentage change in distribution for a given predictor (feature), whereas the CSI measures change in the distribution of the predicted outcomes. PSI is useful in understanding how a distribution of a given feature is changing – something we have done in the previous sections of this chapter as well, albeit a bit differently as we will soon see. Being limited to one feature, PSI cannot give a summarized view of the overall stability. However, since the predicted outcome is dependent on all the features, CSI serves as an indirect measure of how the entire feature set is changing and can be a powerful tool in your toolkit.

That does not mean that PSI is not useful. For example, we have encountered numerous cases during COVID where the models that used macro-economic and market data to predict an outcome failed to maintain their accuracy as the underlying data distribution saw a major unprecedented shift. In the above example, PSI can be used to compute drifts among latent feature that are not used in the model directly but can indirectly impact other features that are being used in the model. The very ability to gauge movement in the stability of a distribution makes it powerful enough to detect any degradation in model performance. Another advantage is that it can work for any kind of data as long as the data can be binned which also makes its less sensitive to outliers.

Similarly, CSI can help us identify any overall model decay without waiting for ground-truth data being available. Since the ground -truth data can take a while to be available, this helps us identify model decay much faster than otherwise possible.

Here's a quick summary of the steps that can be followed to calculate either of the two metrices. The code to cover these steps is included after this:

- Get the data (preferably training data) that would be used as a reference point and the live data.
- Divide the numerical data into bins, preferably deciles (ten bins of equal size).
- Categorical features can be used as-is.
- For each bin, calculate the frequency of observations and % of observations for both training and live data.
- For each bin, take the difference of percentage of observation between training and live data.

- For each bin, take the ratio of percentage of observation between training and live data.
- For each bin, multiply the value obtained in step 5 with natural log of value obtained in step 6.
- Sum for all bins to get the stability index.

Mathematically, given two population sets P and Q (indicating the trained data and the live data, respectively), the stability index can be defined using Kullback-Leibler divergence (D_{KL}). To over the issue of its being non-symmetric, the sum of K-L divergence $[D_{KL}(P \parallel Q) + D_{KL}(Q \parallel P)]$ is used as the base metric.

$$SI(P,Q) = D_{KL}(P \parallel Q) + D_{KL}(Q \parallel P) = \sum_i (P_i) \times \log \frac{P_i}{Q_i} + \sum_i (Q_i) \times \log \frac{Q_i}{P_i}$$

$$\Rightarrow SI(P,Q) = \sum_i (P_i) \times \log \frac{P_i}{Q_i} - \sum_i (Q_i) \times \log \frac{P_i}{Q_i}$$

$$\Rightarrow SI(P,Q) = \sum_i (P_i - Q_i) \times \log \frac{P_i}{Q_i}$$

In addition to the benefits, we have covered so far, the other advantages of this method are:

- Any change in the distribution in any of the bin will lead to increase in overall score irrespective of the direction of the change.
- The logarithm term in stability index calculation has an impact of normalizing the value, which means that if we have a bin with small number of values but a larger deviation, that will make a notable impact on the final score instead of the small bin size leading to little or no impact.
- The PSI/ CSI is symmetric – which means that $SI(P, Q) = SI(Q, P)$.

Let's look at the code for calculating the stability index.

```
def bin_counts(data_train_var, data_live_var, bins):
    # binning with pd.cut
    com_list = list(data_train_var) + list(data_live_var)
    df = pd.DataFrame({'Combined list':com_list})
    df['Rank'] = df['Combined list'].rank()
    df['After cuts'] = pd.cut(df['Rank'].ravel(), bins)

    master_list = list(np.repeat('Actual', len(data_train_var))) + list(np.repeat('Expected', len(data_live_var)))
    binned_df = pd.crosstab(df['After cuts'], pd.Series(master_list)).reset_index()

    actual_counts = binned_df['Actual']
    expected_counts = binned_df['Expected']

    return actual_counts, expected_counts

def stability_index(actual, expected):
    if (actual == 0 or expected == 0):
        return 0
    else:
        return ((actual - expected) * np.log(actual/expected))

def calc_stability_index(data_train_var, data_live_var, bins):
    actual_counts, expected_counts = bin_counts(data_train_var, data_live_var, bins)

    actual_percents = actual_counts/np.sum(actual_counts)
    expected_percents = expected_counts/np.sum(expected_counts)

    si_value = np.sum(stability_index(actual_percents[i], expected_percents[i]) for i
                      in range(0, len(expected_percents)))

    return(si_value)
```

Let's use the same data and find out how PSI or CSI looks like (Fig. 7.7).

For PSI on "Applied Amount"

The plot in Fig. 7.8 shows multiple instances of high drifts as measured using PSI for "Applied Amount" indicating significant in-stability in data distribution calling for further investigation.

Let's jump to CSI for Applied Amount, and see if there is any drift in the predicted probability too.

The plot in Fig. 7.9 shows a few instances of high drifts followed by no drift and, then again, a few instances of high drift as measured using CSI for predicted probabilities indicating significant in-stability in prediction of the model.

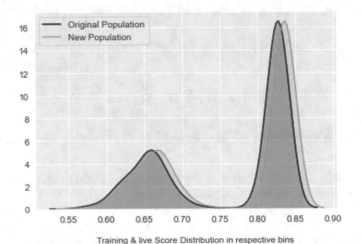

Training & live Score Distribution in respective bins

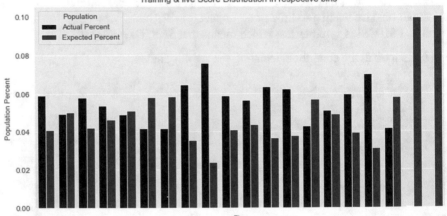

Fig. 7.7 (top to bottom) Training data distribution vs a single live data distribution for "Applied Amount"; population percentage for each bin for training and live data

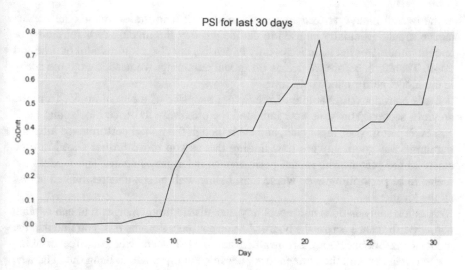

Fig. 7.8 PSI showing population drift for "Applied Amount" during the last 30 days along with high and medium threshold indicator

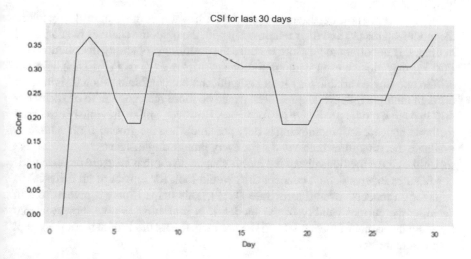

Fig. 7.9 CSI showing population drift for predicted probabilities during last 30 days along with high and medium threshold indicator

Concept Drift

So far, we have seen methods for detecting data drift that aim to detect drifts for either the independent variables (predictors) or the target variable (prediction or outcome). Any drift in a single independent variable may or may not always cause drifts in the final predictions, and hence we need to analyse for drift in multiple predictors. However, the analysis would still be incomplete without detecting drifts in the overall model performance. Any kind of drift in the overall model performance should eventually be used to decide whether the model needs retraining or re-calibration. This decision is not only expensive but also requires a lot of time and effort. Therefore, before we decide on model retraining, we need to note the decay in the model performance over time.

Let's consider our example of predicting the price of a cup of coffee from the previous section. You have now included the proximity to office blocks and businesses as part of your input data, and even though the model performance after the retraining was good, you are now finding that as you move further ahead in time closer to the present, the model performance deteriorates again. The model now seems to be performing even worse than before with prices underestimated across all the locations.

A quick analysis does not reveal any data distribution changes that can explain poorer performance across the board. As you continue the analysis, you find that the model's performance took a downturn around the years where coffee drinking became trendier, and the coffeehouses became the hip place to hang out. The new social standing allowed coffee shops to charge more, this was also coupled with an average coffee shop becoming larger to accommodate more patrons and hence higher real-estate costs.

What you have now experienced is concept drift – where the relationship between the input features (X) and their relationship with the outcome changes over a period of time. It is important to measure because the posterior probability distribution (i.e. $P(X|Y)$) may change even without significant drift in $P(X)$ or $P(Y)$. Thus, a minor drift in one single variable may look insignificant, but combining such insignificant drifts in multiple independent features would be more than enough to create major drift in overall model predictive performance. However, unlike previously discussed methods, concept drift monitoring is only possible when the ground-truth values are available, i.e. it requires both y and $\hat{y}$ for every monitoring instances.

Unlike all the methods discussed in this chapter so far that measure drifts in various features independently, concept drift would look for impact of all drifts taken together on models' overall error rate ($y - \hat{y}$). This helps it play a pivotal role in deciding the current validity of the model. It is important to note, however, that based on the nature of the model and the use case, the methods of detecting concept drift would differ.

Kolmogorov–Smirnov Test

Before we jump into more comprehensive drift detection techniques, we would like to introduce a metric that is widely used in financial services and is also highly recommended by regulators. KS stats or Kolmogorov-Smirnov divergence is one of those metrics which is used to determine the ability of a model to distinguish between events and non-events, by comparing the two cumulative distributions (event and non-event) and returns the maximum difference (distance) between them. Though it does not make it an apt metric to decide or monitor concept drift in itself, this would definitely work as a preliminary indicator of a model performance in terms of predictive quality.

As stated above, KS divergence is theoretically defined as the maximum distance between the distributions of events and non-events, where a wider distance indicates the model's better ability to distinguish between two class (events and non-events or favourable and unfavourable). The steps to calculate KS divergence are as follows:

- Divide the predicted probability into deciles with the highest probability falling into the first decile.
- For each decile, calculate cumulative % of events and non-events followed by the difference between the cumulative distributions.
- KS divergence is the highest difference value.

For monitoring purpose, the model should have similar KS divergence for actual and predicted distributions. The p-value, returned by the KS stats, of more than 0.1 is statistically significant and indicates that the model performance needs to be investigated. The p-value should be less than 0.01 to reject the null hypothesis and conclude that distribution of events and non-events is different).

```
from scipy.stats import ks_2samp
ks_2samp(data.loc[data.y==0,"p"], data.loc[data.y==1,"p"])

Ks_2sampResult(statistic=0.3148725491589702, pvalue=6.83529217017379e-289)
```

In the plot shown in Fig. 7.10, the KS divergence shows the fluctuating KS stats that are significantly different from the training KS stats (the red line). The KS stats values for both the training and the live data also have significant value. However,

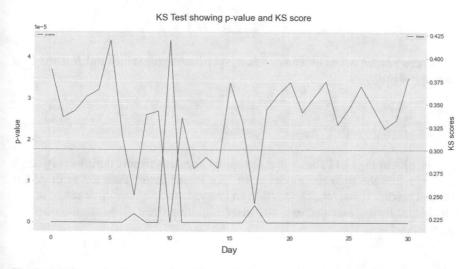

Fig. 7.10 KS test showing model's ability to distinguish between events and non-events. The black line indicates KS distance with red as threshold obtained from KS distance during model training. Given that all p-values (indicated by blue line) are below 0.01, it can be concluded that model outcome's distributions of events and non-events are different

the p-value is low enough to not raise any alarm. The fluctuation should be tested further to check if it is under the predefined threshold and statistically not very different.

Brier Score

So far, all the measures we have talked about utilize the feature space to detect any drift. KS divergence may talk about the model's ability to distinguish between events and non-events but would give little idea on model's accuracy. Since model accuracy is one of the primary metrics during model selection, it would only make sense for us to talk about monitoring it as part of the overall model monitoring. Brier score is a simple metric that monitors the overall accuracy of the model in order to detect its decay over time. Because of its simplicity and intuitive approach, it is also one of the metrics recommended in the Basel norms.

Brier score is well suited for monitoring any kind of classification algorithm which returns prediction probability. As we will see below, it is simply the mean squared error between predicted probabilities and the actual outcome. The error score is always between 0.0 and 1.0 with a higher score indicating higher error and thus high model decay, and the bounded values for the score make it easy for us to define thresholds and compare performance across models. Since we are squaring the difference, the direction of the departure from the actual labels does not matter, just the magnitude does – this means that the prediction probabilities away from actual labels would always lead to a higher error rate.

$$BS = \frac{1}{N}\sum_{t=1}^{N}\left(y-\hat{y}\right)^2$$

Where y is the actual outcome, $\hat{y}$ is the predicted probability and N is the total number of instances.

```
from sklearn.metrics import brier_score_loss

loss = brier_score_loss(y_test, pred_0)
```

The plot in Fig. 7.11 shows that the model accuracy differs during every batch run. Most of the time the model error rate seems lower than the error of the trained model. However, it is advised to keep monitoring the error rates for any significant increase or decrease that may require a deep dive into the cause of such behaviour.

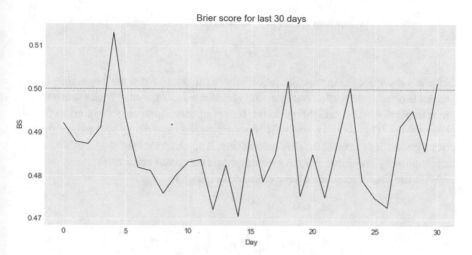

Fig. 7.11 Brier score for last 30 days with threshold indicating Brier score for training data. The plot shows three incidents of model decay breaching threshold set as per the trained model's Brier score

Page-Hinkley Test (PHT)

The Page-Hinkley Test (PHT) is one of the most used recent concept drift detectors that can be used to detect drift in the error for both binary classifier and regression models. It is based on the error rates ($y - \hat{y}$ of the model) and their mean till the current moment. This allows it to retain the memory of recent drift incidents when calculating the drift at any given time. The other measures considered so far calculate the drift at the given instant of the time with no memory of what has happened in the past. This makes PHT unique and desirable when you want to have past incidents of drift reflected in your calculation for the current moment.

For PHT, with decrease in accuracy ($y - \hat{y}$), we would expect the average error rate (ϵ_t) leading up to the current moment (t) to increase in the value. To be able to compute PHT, the cumulative error rate (L_T or m_t) and the minimum error rate (M_T) are computed, where higher L_T corresponds to a higher error rate compared to previous error rate. The PHT then returns an alert when the difference between L_T and M_T is above a specified threshold.

The below equation shows how to compute PHT (PH_T) to identify changes in a given distribution (or error distribution).

$$m_t = \sum_{t=1}^{T} \left(\epsilon_t - \overline{\epsilon_T} - \alpha \right)$$

where ϵ_t is error value at time t;

$\overline{\epsilon_T}$ is current mean calculated as $\dfrac{1}{T} \sum_{t=1}^{T} \epsilon_t$

α is tolerance for change

$$\text{Minimum value}\left(M_T\right) = \min\left(m_t, \text{where } t = 1...T\right)$$

$$PH_T = m_T - M_T$$

Since it's a sequential error calculation where all errors from the beginning till the current time are taken into consideration (from, $t = 1...T$), it may make sense to give more importance and weightage to the recent observations as compared to the old observations. This would eventually incorporate the time decay and ensure that old error rate has less impact on PH_T calculation. The observations are weighted as per their topicality and force PH_T to give emphasis on recent error rates.

With the time decay, the above equations change to

$$L_0 = 0$$

$$L_T = m_t = \frac{T-1}{T}L_{T-1} + \left(\epsilon_T - \overline{\epsilon_T} - \alpha\right)$$

$$M_T = \min\left(L_t, \text{where } t = 1...T\right)$$

$$PH_T = L_T - M_T$$

In the above equation, the time decay is introduced in the calculation by multiplying L_{T-1} by with $\frac{T-1}{T}$. More recent samples will lead to higher $\frac{T-1}{T}$ value, thus giving more weight to recent observations.

Once you detect a drift that requires an action, you may want to restart with a clean slate because you have retrained the model. Any information about the previous drifts after the update is not needed and is going to incorrect determination. At this point, you can choose to reset the mean error ($\overline{\epsilon_T}$) and time ($t = 1, 2, ...T$), and the next series of observations are then deemed to start from $T = 0$.

```
for i in range(len(pred_0)):
    ph.add_element(pred_0[i])
    if ph.detected_change():
        print('Change has been detected in data: ' + str(data_stream[i]) + ' - of index: ' + str(i))
```

The plot in Fig. 7.12 shows the error rate for last 30 days and seven instances of major drift as calculated using Page-Hinkley Test. These seven instances of major error should further be compared to error rate of the trained model for better decision on the validity of the model in production.

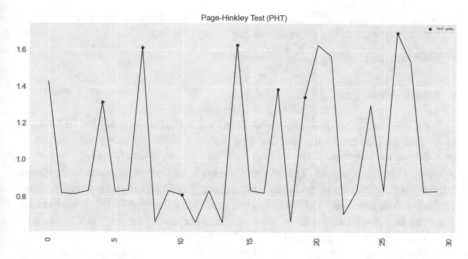

Fig. 7.12 Plot showing PHT test with days marked as red dots when PHT detected drifts

Early Drift Detection Method

Many a time it can be difficult to detect gradual concept drifts. Detecting an abrupt drift is much simpler and can be done using any methods discussed earlier, but a lot of time slow and hidden drift goes undetected, until it is too late to act on it within the time to prevent breaching the thresholds. In such cases, early drift detection method (EDDM) is a useful method to detect concept drift. Unlike other methods, EDDM not just uses the errors $(y - \hat{y})$, but the difference in the two errors. It uses the distance $(\epsilon_t - \epsilon_{t-1})$ between subsequent model errors to detect drifts.

Before continuing with EDDM, let us briefly look into the older version of EDDM known as drift detection method (DDM). The drift detection method (DDM) uses number of errors in the sample given. For each observation, the error rate is the probability p_i with standard deviation s_i calculated as $\sqrt{\dfrac{p_i(1-p_i)}{i}}$ assuming that error rate will decrease with increase in sample size. This method is more suited for binary classification and most of the time captures change class frequencies. The method begins by evaluating p_i and s_i and stores the value of p_i and s_i when $p_i + s_i$ reaches its minimum value. A warning is raised when $p_i + s_i \geq p_{min} + 2s_{min}$, and drift is indicated when $p_i + s_i \geq p_{min} + 3s_{min}$. This approach is good for detecting abrupt changes only.

Coming back to EDDM, as said above, it is based on the distance between two adjacent errors and not on the numbers of error. On the high level, it starts with calculation of average and standard deviation between distances of errors as μ_i' and σ_i', respectively. The maximum values for μ_i' and σ_i' are stored as μ_{max}' and σ_{max}' – representing the instance where the distance between error is maximum. The EDDM scores are calculated as $\dfrac{\mu_i' + 2\sigma_i'}{\mu_{max}' + 2\sigma_{max}'}$. In this case, the warning is raised as per below rules (these thresholds are a good start and should work for most scenarios; however, you may want to use other values for your problem):

- Warn and start caching $\dfrac{\mu'_i + 2\sigma'_i}{\mu'_{max} + 2\sigma'_{max}} \le 0.95$

- Alert and reset max $\dfrac{\mu'_i + 2\sigma'_i}{\mu'_{max} + 2\sigma'_{max}} \le 0.90$

With the underlying math explained, let's look at the intuition behind EDDM. We know that EDDM tracks the average and the max error. By using the ratio of the two, it tries to determine if the swings that lead to a new maximum are increasing relative to the average error. Therefore, if the error rate has been more or less steady and we start seeing new highs in the model error, it indicates that the drift thresholds are likely to be breached. Due to its higher sensitivity than the DDM, it is able to do that even then the change is slow. However, this sensitivity can also make it react to the noise in the data and reduce the usability in that case.

In the code below, we begin by calculating the adjacent errors, followed by the mean and the standard deviations. Then we calculate the values that are needed to compute the EDDM score.

```
def find_eddm_max_index(df, metric, max_metric):
    return df[df[metric] == max_metric].index[0]

def create_eddm_df(error):
    df = pd.DataFrame({'Error':error})
    df['DiffError'] = df['Error'].diff()
    df['pi'] = df['DiffError'].rolling(window=len(error), min_periods=1).mean()
    df['si'] = df['DiffError'].rolling(window=len(error), min_periods=1).std(ddof=0)
    df['pi+2si'] = df['pi']+(2*df['si'])
    df['max_pi+2si'] = df['pi+2si'].rolling(window=len(error), min_periods=1).max()

    df = df.fillna(0)

    df['index_max_pi+2i'] = df['max_pi+2si'].map(lambda x:find_eddm_max_index(df, 'pi+2si', x))
    df['pi_max'] = df['pi'][df['index_max_pi+2i']].to_numpy()
    df['si_max'] = df['si'][df['index_max_pi+2i']].to_numpy()

    return df
```

In the code below, EDDM is generated by taking the ratio $\dfrac{\mu'_i + 2\sigma'_i}{\mu'_{max} + 2\sigma'_{max}}$ which is then used to generate warning as per the threshold declared.

```
def check_eddm_threshold(eddm_signal):
    eddm_warning_thresh = 0.95
    eddm_drift_thresh = 0.90

    eddm_warning_boolean = eddm_signal < eddm_warning_thresh
    eddm_drift_boolean = eddm_signal < eddm_drift_thresh

    return eddm_warning_boolean, eddm_drift_boolean

#main function
def eddm_warning(error, time):
    df = create_eddm_df(error)
    df['Dates'] = time
    df['EDDM_signal'] = df['pi+2si']/(df['pi_max']+(2*df['si_max']))
    eddm_warning_thresh, eddm_drift_thresh = check_eddm_threshold(df['EDDM_signal'])
    df['EDDM_Warn'] = eddm_warning_thresh
    df['EDDM_Drift'] = eddm_drift_thresh

    return df
```

Using the above set of code, a data frame is generated that has the actual error $(y - \hat{y})$, the difference between two adjacent error $(\epsilon_t\text{-}\ \epsilon_{t-1})$, mean and standard deviation as μ'_i and σ'_i, the max value of $\mu'_{max} + 2\sigma'_{max}$ till the current instance and the

	Error	DiffError	pi	si	pi+2si	max_pi+2si	index_max_pi+2i	pi_max	si_max	Dates	EDDM_signal	EDDM_Warn	EDDM_Drift
0	0.7663	0.0000	0.000000	0.000000	0.000000	0.0000	0	0.0000	0.0000	0	NaN	False	False
1	0.6288	-0.1375	-0.137500	0.000000	-0.137500	-0.1375	1	-0.1375	0.0000	1	1.000000	False	False
2	0.7201	0.0913	-0.023100	0.114400	0.205700	0.2057	2	-0.0231	0.1144	2	1.000000	False	False
3	0.6517	-0.0684	-0.038200	0.095817	0.153434	0.2057	2	-0.0231	0.1144	3	0.745913	True	True
4	0.6268	-0.0249	-0.034875	0.083180	0.131484	0.2057	2	-0.0231	0.1144	4	0.639205	True	True
5	0.6204	-0.0064	-0.029180	0.075265	0.121350	0.2057	2	-0.0231	0.1144	5	0.589937	True	True
6	0.7613	0.1409	-0.000833	0.093479	0.186125	0.2057	2	-0.0231	0.1144	6	0.904837	True	False
7	0.6881	-0.0732	-0.011171	0.090174	0.169176	0.2057	2	-0.0231	0.1144	7	0.822439	True	True
8	0.7296	0.0415	-0.004587	0.086130	0.167672	0.2057	2	-0.0231	0.1144	8	0.815127	True	True
9	0.6713	-0.0583	-0.010556	0.082940	0.155324	0.2057	2	-0.0231	0.1144	9	0.755098	True	True

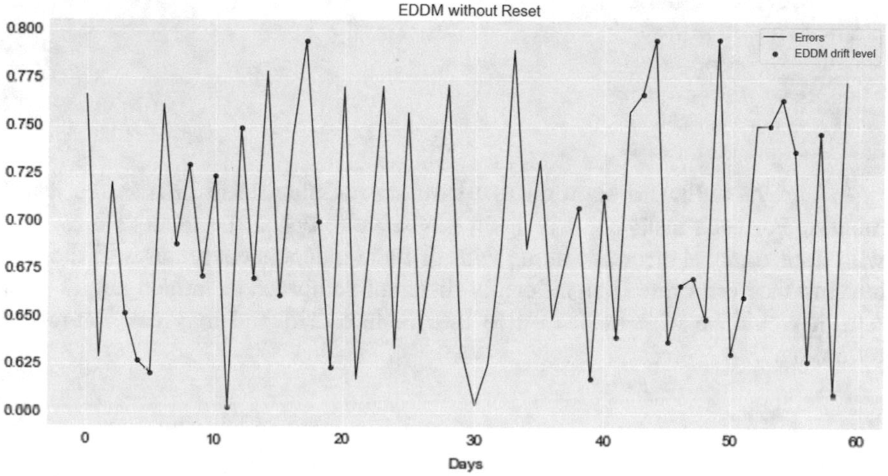

Fig. 7.13 EDDM for error from a regression model before re-initialization

EDDM signal calculated as the ratio $\dfrac{\mu_i' + 2\sigma_i'}{\mu_{max}' + 2\sigma_{max}'}$. Figure 7.13 shows EDDM for error from a regression model before re-initialization.

Extending the concept, the loop in the following code example checks for EDDM signal to reach the warning threshold and re-initializes (reset) the entire function. The reset assumes that you are taking some action to address the drift before re-initializing the entire function. A reset without addressing drift will mean that the drift still exists, we are just not identifying it anymore.

```
df = eddm_warning(data['Error'], data['Days'])

drift_time = []
drift_error = []

for i in range(0, len(df)):
    if (df['EDDM_Drift'][i] == True):
        error = data['Error'].iloc[i:,]
        time = data['Days'].iloc[i:,]
        df = eddm_warning(error, time)

        drift_time.append(data['Days'].iloc[i])
        drift_error.append(data['Error'].iloc[i])

        print("{}. Drift time: {}".format(i, data['Days'].iloc[i]))
        print("{}. Drift error: {}".format(i, data['Error'].iloc[i]))
```

```
3.  Drift time: 3
3.  Drift error: 0.3095982162480306
7.  Drift time: 7
7.  Drift error: 0.3095982162480306
10. Drift time: 10
10. Drift error: 0.3095982162480306
16. Drift time: 16
16. Drift error: 0.46516645233016246
24. Drift time: 24
24. Drift error: 0.3095982162480306
28. Drift time: 28
28. Drift error: 0.3095982162480306
```

In Fig. 7.14, EDDM shows a few instances of significant drift in the last 2 months. For each instance, it is again advisable to check the model's error rate with trained model error rate along with drifts in independent features. If the current instance error rate is significantly different compared to trained model error rate, it would be wise to deep dive into feature drift and may call for model retraining.

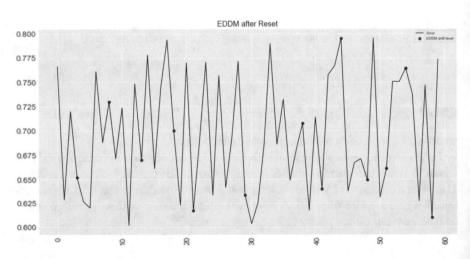

Fig. 7.14 EDDM for error from a regression model after re-initialization when drift is detected

Hierarchical Linear Four Rate (HLFR)

The techniques so far in the chapter have focussed on either error rate or the accuracy of the model. In a few cases, simply finding the drift in the overall error rate or the accuracy won't give the right picture. For instance, in an imbalance classification problem, it is AUC (based on confusion matrix) which is more important and not simply accuracy measures. Similarly, in a few business cases, it would be important to measure a particular accuracy metric and not overall error rate. For instance, a business case might want to reduce the false positive rate (FPR) and won't worry much about the overall accuracy decay. Hierarchical linear four rate (HLFR) is based on the error monitoring of metrics derived from confusion matrix. There's four in its name because of the four metrics it focuses on – true positive rate (TPR), true negative rate (TNR), positive predictive value (PPV) and negative predictive value (NPV). It's important to note here that this method is suitable for imbalanced class problem and only those cases where metrics using confusion matrix can be generated.

$$TPR = \frac{TP}{TP + FN}$$

$$TNR = \frac{TN}{TN \mid FP}$$

$$PPV = \frac{TP}{TP + FP}$$

$$NPV = \frac{TN}{TN + FN}$$

The fundamental concept of this method is that any change in data will lead to change in any or all of this matrix as it will impact accuracy of the model which will create ripples in the confusion matrix. To calculate the drift, the change in the confusion matrix is considered rather than recalculation of confusion matrix from the scratch. It is based on incremental update of the chosen metric. To calculate the change in the metrics, R_t^* is computed which is a linear combination of past and present value of the metric.

$$R_t^* \leftarrow \eta^* R_{t-1}^* + \left(1 - \eta^*\right) \times 1_{y = \hat{y}}$$

where current rate is R_t^* and previous rate $R_{t-1}^* \in (TPR, TNR, PPV, NPV)$
η^* is time decay where $0 < \eta^* < 1$

R_t^* is updated incrementally using linear combination of past value and time decay factor η^*. Interestingly, the time decay factor is only considered in calculation when the predicted outcome is equal to the actual outcome ($y = \hat{y}$). This very condition is to stress upon the topicality and accuracy of the model.

The time decay factor in calculation of R_t^* is able to suggest more about the current model metric performance during change in concept drift along with including the past drift (although with less weight) in the calculation. A larger time decay (η^*) value will allow this method to learn more about recent drifts rather than giving same weight to drifts of past.

This time-decaying factor η^* is made dynamic using weights where a higher weight is assigned when the current metric value is higher than the previous and vice versa.

$$\eta^* = \begin{cases} \left(\eta^*_{t-1} - 1\right)e^{-\left(R^*_t - R^*_{t-1}\right)} + 1 \ when \ R^*_t \geq R^*_{t-1} \\ \left(1-\eta^*_{t-1}\right)e^{\left(R^*_t - R^*_{t-1}\right)} + \left(2\eta^*_{t-1} - 1\right) when \ R^*_t < R^*_{t-1} \end{cases}$$

Illustration:

	t	TN	FP	FN	TP	Y	Yhat	Rhat	Decay n	metricvalue_1_2	Y_Yhat
0	0	1.0	2.0	1.0	1.0	1	0	0.5000	0.9000	False	False
1	1	1.0	2.0	1.0	2.0	1	1	0.4500	0.8951	False	True
2	2	2.0	2.0	1.0	2.0	0	0	0.5077	0.9010	True	True
3	3	2.0	2.0	1.0	3.0	1	1	0.5077	0.9010	False	True
4	4	3.0	2.0	1.0	3.0	0	0	0.5564	0.9057	True	True
5	5	3.0	2.0	1.0	4.0	1	1	0.5564	0.9057	False	True
t	TN	FP	FN	TP	Y	Yhat	Rhat	Decay n	metricvalue_1_2	Y_Yhat	
25	25	9.0	5.0	4.0	12.0	1	1	0.6756	0.9170	False	True
26	26	9.0	5.0	4.0	13.0	1	1	0.7025	0.9192	False	True
27	27	9.0	6.0	4.0	13.0	1	0	0.7266	0.9212	False	False
28	28	10.0	6.0	4.0	13.0	0	0	0.6693	0.9168	True	True
29	29	11.0	6.0	4.0	13.0	0	0	0.6693	0.9168	True	True

For t = 4,

$$0.9010 * 0.5077 + \left(1 - 0.9010\right) * 1 = 0.5564$$

$$as \ y == \hat{y}; R^*_t = \eta^*_{t-1} \times R^*_{t-1} + \left(1 - \eta^*_{t-1}\right) \times 1$$

For t = 28,

$$0.9212 * 0.7266 + \left(1 - 0.9212\right) * 0 = 0.6693$$

$$as \ y \neq \hat{y}; R^*_t = \eta^*_{t-1} \times R^*_{t-1} + \left(1 - \eta^*_{t-1}\right) \times 0$$

The code below defines the four metrics we want to monitor. You may choose the use as many or as few metrics as what make sense for your requirements. The approach defined in this section can also be used for any other metric, and the choice is not really limited to the four described here (see the note at the end of this section). The function would return the computed value as per the definition of the metric defined.

```
def compute_tpr(confusion_matrix):
    return confusion_matrix.tp / (confusion_matrix.tp + confusion_matrix.fn)

def compute_tnr(confusion_matrix):
    return confusion_matrix.tn / (confusion_matrix.tn + confusion_matrix.fp)

def compute_ppv(confusion_matrix):
    return confusion_matrix.tp / (confusion_matrix.tp + confusion_matrix.fp)

def compute_npv(confusion_matrix):
    return confusion_matrix.tn / (confusion_matrix.tn + confusion_matrix.fn)

METRICS_FUNCTION_MAPPING = {
    'tpr': compute_tpr,
    'tnr': compute_tnr,
    'ppv': compute_ppv,
    'npv': compute_npv
}
```

In the code below, the initial value is declared followed by update_metric function which is to update R_t^* with previous values and weighted η_*. The update_metric function calls update_decay function.

In update_decay, two incremental updates are computed. If the current metric (t) value (of the four defined above) is not equal to previous value ($t-1$), R_t^* is updated using a weighted function comprising the last decay and R_t^* and a direct accuracy check, else if the current and previous metric value are same, R_t^* remains as-is. In the next step in the same function, time decay component is updated as η_* is calculated as per the equation and conditions described above based on consecutive R_t^* values.

```
class PerformanceMetric(object):
    def __init__(self, metric_name, decay):

        self.metric_name = metric_name
        self.decay = decay #[0.7]
        self.metric_value = [0.5]
        self.R = [0.5]

    def reset_internals(self):
        self.decay = decay #[0.7]
        self._R[-1] = 0.5
        self._P[-1] = 0.5

    def update_metric(self, confusion_matrix, y_actual, y_pred):

        self.metric_value.append(METRICS_FUNCTION_MAPPING[self.metric_name](confusion_matrix))
        r_hat, decay_t = self.update_decay(y_actual, y_pred)

        return r_hat, decay_t

    def update_decay(self, y_actual, y_pred):
        if (abs(self.metric_value[-1] - self.metric_value[-2]) > 0):
            self.R.append(self.decay[-1] * self.R[-1] + (1 - self.decay[-1]) * int(y_actual == y_pred))
        else:
            self.R.append(self.R[-1])

        if (self.R[-1] < self.R[-2]):
            self.decay.append((1 - self.decay[-1])*np.exp(self.R[-1] - self.R[-2]) + (2*self.decay[-1] - 1))
        else:
            self.decay.append((self.decay[-1] - 1)*np.exp(-(self.R[-1] - self.R[-2])) + 1)

        return self.R, self.decay
```

In the next set of steps, we loop through the data to calculate all the required metrics.

```
metric_choice = 'tpr'
decay = [0.9]
metrics = {metric_choice: PerformanceMetric(metric_choice, decay)}
confusion_matrix = StreamingConfusionMatrix()
```

```
import pandas as pd

b=[]
data = pd.DataFrame([])
#decay_t = [0.9]

for t, (y, y_hat) in enumerate(zip(y_true[:30], y_pred[:30])):

    confusion_matrix.update_confusion_matrix(y, y_hat)


    for metric in metrics.values():
        r_hat, decay_t = metric.update_metric(confusion_matrix, y, y_hat)
```

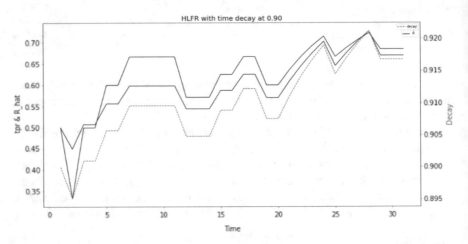

Fig. 7.15 The plot showing metric value, decay and time decay using HLFR with time decay set at 0.90 and accuracy metric being considered is TPR

The plot generated (Fig. 7.15) shows increasing trend of drift as measured by HLFR for TPR. However, it is interesting to see increasing value of TPR but that also indicates that there is a need to check drift among other accuracy rates too.

Note: Monitoring Fairness Metrics - HLnR

All the techniques we have covered in this chapter are related to monitoring the model performance or other measures of performance. Monitoring them gives us the confidence that we can keep the model true to the purpose and hence as a result have it perform optimally not just for the accuracy but also for the fairness.

We have covered metrics as either for monitoring data drift or concept drift, but we might have as well classified them as the model monitoring metrics that do not require the ground truth vs the metrics that require the ground truth. The metrics themselves can be broken down into what is being measured and how is it being measured – with the mechanism of measurement changing between the different metrics even though the parameters measured may not change.

With this in mind, we want to finish this chapter with the concept of monitoring fairness metrics. Let's consider HLFR; it has a threshold defined for the warning and another for indicating a need for action, followed by a reset. The parameters that HLFR monitors are based on the confusion matrix, just the way most of the fairness metrics are. Therefore, we would propose using the HLFR method to monitor fairness metrics like equality of odds, equal opportunity and predictive parity. Simply by replacing (or including in addition to) the four model accuracy metrics with the fairness metric HLFR can be used as HLnR where n is the number of metrics being monitored – be it other accuracy metric or fairness metrics.

To extend this approach further, it might also be useful for you to monitor additive counterfactual fairness (ACF) metrics (as well as counterfactual unfairness or CUF metrics) using EDDM or PHT methods for monitoring drift in model fairness for regression-based models. As we have seen earlier that in case of regression-based use cases, fairness metric is defined as CUF (counterfactual unfairness) based on prediction errors or the difference in prediction between two groups of a protected features are used to estimate the discrimination caused by the model. Thus, these metrics can be used as errors and can help monitor deviation in discrimination as used in EDDM, PHT or any other techniques discussed above that is well-suited for non-binary error types.

Conclusion

We know that all models deteriorate in performance over time due to changing distributions or drifts. In this chapter, we have seen the two types of drifts and some of the measures that we can utilize to detect drifts as early as possible. Just as setting up the model monitoring for the key metrics is important, we also need to have a good intuition of what these metrics and changes in them mean for the model and the impact they can have on the quality of the prediction and on the users as a result. This understanding allows us to identify thresholds for changes in the selected metrics to generate a warning and set alert levels that we cannot afford to exceed. Together, the metrics, thresholds and the measures to track the metrics help us set up the model monitoring.

In this chapter, we have covered the techniques and the metrics that should be useful for most of the scenarios and problems. However, the model monitoring is not limited to these techniques or metrics. You may want to try and investigate use of metrics that are more relevant to the problems you are solving – this may even mean that you create custom metrics that are the best fit for your solution and then use a technique for detecting any drift in the chosen metric.

We would, however, recommend starting with simple metrics before delving deep into more complex or custom ones because they come with simpler intuition and are easy to explain. Another pitfall to avoid is spending a lot of time in deciding the metric but not as much attention to fine tuning the threshold levels. You want to raise a warning in time so that if an action has to be taken – including any retraining – you are able to take that before the breach thresholds for the metrics are violated. Sometimes an outlier event may mean that a breach is unavoidable, like we've all experienced with the global pandemic, but a well-planned threshold level should still help us prevent a breach most of the times.

Finally, any monitoring you set up is as good as the action you take when the alarms are raised by it. Before you decide to retrain/replace the model, it would be useful to begin by determining if it's a real or a false alarm. Checking other metrics to establish patterns would provide you better evidence for the drift and also help decide the approach ahead. Once you've confirmed that you are experiencing a drift and the pace/magnitude of it, you can make an informed decision on how to proceed forward.

Just remember that the model's development lifecycle doesn't stop with model monitoring. This is where it really begins.

Bibliography

Baena-Garcia, M. et al. (no date) Early drift detection method ?, Upc.edu. Available at: https://www.cs.upc.edu/~abifet/EDDM.pdf (Accessed: April 16, 2021).

Bellemare, M. G. et al. (2017) "The Cramer Distance as a Solution to Biased Wasserstein Gradients," arXiv [cs.LG]. Available at: http://arxiv.org/abs/1705.10743.

Bhalla, D. (no date) Calculate KS Statistic with Python, Listendata.com. Available at: https://www.listendata.com/2019/07/KS-Statistics-Python.html (Accessed: April 16, 2021).

Brownlee, J. (2018) A gentle introduction to probability scoring methods in Python, Machinelearningmastery.com. Available at: https://machinelearningmastery.com/how-to-score-probability-predictions-in-python/ (Accessed: April 16, 2021).

Core.ac.uk. 2021. [online] Available at: <https://core.ac.uk/download/pdf/11581818.pdf> [Accessed 16 April 2021].

Elliott, M. (2017) What a cup of coffee cost each president since 1920, Cheatsheet.com. Available at: https://www.cheatsheet.com/money-career/historical-look-at-the-cost-of-a-cup-of-coffee.html/ (Accessed: April 16, 2021).

Facom.ufu.br. 2021. [online] Available at: <http://www.facom.ufu.br/~elaine/disc/MFCD/change.pdf> [Accessed 16 April 2021].

GitHub. 2021. alipsgh/tornado. [online] Available at: <https://github.com/alipsgh/tornado/blob/master/drift_detection/eddm.py> [Accessed 16 April 2021].

GitHub. 2021. ModelOriented/drifter. [online] Available at: <https://github.com/ModelOriented/drifter> [Accessed 16 April 2021].

GitHub. 2021. pranab/beymani. [online] Available at: <https://github.com/pranab/beymani/blob/366a6909a745ece4c0cbefeb4511331509bb6c6c/python/lib/sucodr.py> [Accessed 16 April 2021].

GitHub. 2021. scikit-multiflow/scikit-multiflow. [online] Available at: <https://github.com/scikit-multiflow/scikit-multiflow/blob/a7e316d/src/skmultiflow/drift_detection/eddm.py#L6> [Accessed 16 April 2021].

GitHub. 2021. scikit-multiflow/scikit-multiflow. [online] Available at: <https://github.com/scikit-multiflow/scikit-multiflow/blob/a7e316d/src/skmultiflow/drift_detection/page_hinkley.py#L4> [Accessed 16 April 2021].

GitHub. 2021. thuijskens/hlfr. [online] Available at: <https://github.com/thuijskens/hlfr/blob/1dce4dff918ce1a0fdc3b78c9794688ef133e3e9/hlfr/hlfr.py> [Accessed 16 April 2021].

II. Mouss, D. Mouss, N. Mouss and L. Sefouhi, "Test of Page-Hinckley, an approach for fault detection in an agro-alimentary production system," 2004 5th Asian Control Conference (IEEE Cat. No.04EX904), Melbourne, VIC, Australia, 2004, pp. 815-818 Vol.2.

Kang, Y. and Zadorozhny, V. (2018) "Process monitoring using maximum sequence divergence," arXiv [stat.ML]. Available at: http://arxiv.org/abs/1807.03387.

Kiritz, N. and Sarfati, P. (2018) "Supervisory guidance on model risk management (SR 11-7) versus enterprise-wide model risk management for deposit-taking institutions (E-23): A detailed comparative analysis," SSRN Electronic Journal. doi: https://doi.org/10.2139/ssrn.3332484

Lin, A. Z., Depot, L. and Foothill Ranch (no date) Examining distributional shifts by using population stability index (PSI) for model validation and diagnosis, Lexjansen.com. Available at: https://www.lexjansen.com/wuss/2017/47_Final_Paper_PDF.pdf (Accessed: April 16, 2021).

Medium. 2021. KI divergence and PSI. [online] Available at: <https://medium.com/@zhangbo-jun_25458/kl-divergence-and-psi-4c43715e69fc> [Accessed 16 April 2021].

Ramdas, A., Garcia, N. and Cuturi, M. (2015) "On Wasserstein two sample testing and related families of nonparametric tests," arXiv [math.ST]. Available at: http://arxiv.org/abs/1509.02237 (Accessed: April 16, 2021).

Sebastião, R. and.Fernandes, J. M. (2017) "Supporting the page-Hinkley test with empirical mode decomposition for change detection," in Lecture Notes in Computer Science. Cham: Springer International Publishing, pp. 492–498.

Sebastião, R. et al. (2013) "Real-time algorithm for changes detection in depth of anesthesia signals," Evolving systems, 4(1), pp. 3–12.

Stat.cmu.edu. 2021. [online] Available at: <https://www.stat.cmu.edu/~larry/=sml/Opt.pdf> [Accessed 16 April 2021].

Sun, C., Azari, N. and Turakhia, C. (2020) "Gallery: A machine learning model management system at Uber." OpenProceedings.org.

Taplin, R. and Hunt, C. (2019) "The population accuracy index: A new measure of population stability for model monitoring," Risks, 7(2), p. 53.

The fed - supervisory letter SR 11-7 on guidance on model risk management -- April 4, 2011 (no date) Federalreserve.gov. Available at: https://www.federalreserve.gov/supervisionreg/srletters/sr1107.htm (Accessed: April 16, 2021).

Yu, S. and Abraham, Z. (2017) "Concept drift detection with hierarchical hypothesis testing," in Proceedings of the 2017 SIAM International Conference on Data Mining. Philadelphia, PA: Society for Industrial and Applied Mathematics, pp. 768–776.

Yu, S. et al. (2019) "Concept drift detection and adaptation with hierarchical hypothesis testing," Journal of the Franklin Institute, 356(5), pp. 3187–3215.

Chapter 8
Data and Model Privacy

Introduction

In 2006, Netflix released 100 million anonymized movie ratings data (containing unique subscriber ID, movie title, release year and date of rating) for an online competition to build an algorithm that would predict movie rating by a subscriber. Within 2 weeks of the data being released, two researchers Arvind Narayanan and Vitaly Shmatikov were able to reverse-engineer the so-called anonymized data and identify the subscriber by matching it with other similar data sources like IMDB. So much so, they not only identified the viewing history of the users, but also their apparent political preferences and other potentially sensitive and personal information.

Before we look at different ways to protect private or personal data, let's spend a moment to understand what private data is. GDPR defines personal data (personally identifiable information or PII) as any information relating to an identified or identifiable natural person ("data subject"); an identifiable natural person is one who can be identified, directly or indirectly, in particular by reference to an identifier such as a name, an identification number, location data, an online identifier or to one or more factors specific to the physical, physiological, genetic, mental, economic, cultural or social identity of that natural person.

This includes information shared in the social media (text, images, videos, audios), locations using the user's phone, user's device usage patterns, shopping patterns, transaction behaviours, financial information and other personal information like health records, relationship information or even near future plans.

Few years ago, in 2000, computer scientist Latanya Sweeney combined the anonymized hospital data released by the Massachusetts Group Insurance Commission with the 1990 US census records and other information in public domain to identify the health records of the Governor of Massachusetts. Sweeney estimated that 87% of the US population can be identified uniquely using just three

S. Agarwal, S. Mishra, *Responsible AI*,
https://doi.org/10.1007/978-3-030-76860-7_8

parameters: five-digit zip code, gender and date of birth. This is known as linkage attack where anonymized data is combined with other public datasets using common attributes between the datasets to identify individual information. Once a common approach and considered good enough, removing the basic customer identifiers or obfuscating them with hash keys is now known to be a weak approach that does not really protect the private data on its own. The approaches we will see to protect data in this chapter will aim to make data safe from individual identification while keeping it useful for any statistical analysis.

The risks posed by the data that does not truly protect private data are not limited to potential hacks where the data is accessed for nefarious purposes. As we build machine learning models using this data, it can cause biases to come in or for the model to learn relationships dependent on sensitive features. Finally, there is also the risk of private information being available to your teams working on the data and resulting liabilities.

To mitigate these risks, the approach should be to enhance the privacy of the data as soon as you can in your data pipeline. If the data is made private before it gets into your datastore, any downstream consumption will not be able to access the private information. As we will see soon, adding privacy to the data also reduces bias from the model by marking information that is used during the modelling. This in turn impacts the fairness treatment and the explainability and monitoring too. For example, if a model is trained using geospatial data, health records or demographic data, there is a high probability of bias being present in the data. Adding privacy, at different levels, can help us prevent ourselves from one or more of these challenges.

In addition to the above risks, lack of privacy in the data or the model also leaves your model susceptible to adversarial attacks. The adversarial attacks are classified as white box (where the attacker has the model parameters), black box attacks (where the attacker does not know the model parameters or the architecture) or grey box (somewhere in between the two). In both cases, the objective of the attack can be to:

1. Identify and leak data about the users or model parameters
2. Cause misclassification of the target observation
3. Reduce the model accuracy

We will see later in this chapter that adding privacy to your model gives a reliable protection against such attacks by adding noise to the information.

Note: Adversarial Attacks on ML Models

Linear models are very sensitive to the training data. A change in the training data can skew the model parameters, and this vulnerability can be exploited by an attacker to determine if a record was part of the training data. Similarly, KNN models carry a lot of training information and can be exploited by the attacker if the data does not have privacy built into it.

Typically, in a white box attack, an attacker would aim to dilute the model by disturbing the model parameters and thus forcing the model to give wrong outcomes. While in case of black box attack, attacker sends data (assumed to be from the same distribution) to the target model and collects the output which is then used to create a shadow model which mimics the target model. This kind of attack is feasible when attacker has access to the target model mostly possible during transfer learning or models in the cloud.

In a black box attack, training data is used to find the model parameters, while a white box attack can reveal the training data. For example, using some information about the patient, a model trained on medical data can allow an attacker to access confidential information (model inversion) about patient health conditions. Similarly, membership of an individual in training data can be easily found by sending the data to the model and checking model's prediction confidence. The model's prediction (membership inference) would be more confident if the person was part of the training sample, as compared to when they were not. This is very true in case of model that is used for predicting say credit score by a bank, readmission probability by a hospital or survival probability of cancer patient of a locality.

Basic Techniques

Let's begin with the basic approaches to masking the data. As we saw in the examples earlier, they don't offer protection on their own but can serve as a good data hygiene when you have data with personal information. This can be useful if you need to share your data, especially when used in addition to the techniques discussed later in the chapter.

Hashing

A one-way hashing function is a common way to convert sensitive fields into a fixed length unique hash to protect the data from being linked to a specific user. To avoid hashing clash, a unique salt string can be added to the value to create a unique hash, which is then stored. While this creates one-way unique values, there are several challenges with using hashing for masking sensitive fields.

Hashing is useful when you want to convert a value into a unique value, often implemented as a one-way mechanism to securely store passwords. Most of the problems where you will employ machine learning to solve it will involve high-dimensional data where it is already very unique. This limits the value that hashing is going to provide. The other challenge with hashing is that even though it will mask the fields that are being hashed, it cannot mask the relationships that the unmasked data can reveal. As we go along in the book, we will see how those relationships can reveal biases, and handling them correctly plays an important part in creating a responsible AI model.

Finally, once a field has been hashed, it cannot be used for any exploratory analysis or model training. This restricts the use of the technique to the fields that are unique identifiers of a person and would not have been included in the analysis or the training anyway. An example of such a field is customer ID which is needed as a unique identifier to tell the records separately but using the original value can reveal the individual's identity. Beyond the masking of ID values, we do not recommend using hashing.

K-Anonymity, L-Diversity and T-Closeness

There are plenty of techniques for data anonymization. A few of them are simple masking, data aggregation, data grouping or converting numeric data to categories. However, most of these methods are quite weak and lead to a lot of information loss. These techniques also don't prevent attackers from reverse engineering to find out the sensitive information. Let's quickly go through some techniques that are stronger and don't require us to lose a lot of data.

k-anonymity is a technique to ensure that the information of a single person in the data would be similar to k-1 individual in the dataset. For instance, if the attacker would like to identify an individual using some of their private data like postcode, marital status and age, then they would end up finding K individuals with same combination of protected features. It would further generalize some of features (e.g. convert all date of birth between 1980 and 1989 to 1980) or use suppression for anonymity like reducing a six-digit postcode to first three digits followed by '*' (CV1*, WG1*, etc.). However, if the attacker is able to access multiple datasets with different k-anonymity, they can use that to identify the missing data and ultimately can reveal information that is private.

l-diversity builds upon k-anonymity and creates intra-group diversity for sensitive features. However, if the attacker has the knowledge of the distribution of the original data, l-diversity can be decoded to find out original information. Building on the drawback of l-diversity, t-closeness is computed as the distance between the sensitive attribute of a class and the complete data set which is less than or equals to t. The distance between both the distribution is computed using Wasserstein distance (as seen in the previous chapter).

Differential Privacy

Let's consider an example – you are working on using the data from calls to the emergency services to understand if there are any local trends that may require attention. The data contains the information that the callers provided to the emergency dispatcher (including name, address, age, symptoms) and the diagnosis and the details of the subsequent treatment given. Removing fields from the dataset is not only going to increase privacy – just the street name and the age might identify the caller uniquely – but will also render the data unusable with a lot of missing information.

Differential privacy helps us overcome both these challenges. Put simply, differential privacy works by introducing noise to the data so that it does not really significantly change the distribution, keeping the data usable for any analysis and model training, but distorting it enough to prevent specific individuals to be identified. For our example above, changing some of the fields in the dataset, say the age and the street name, to values in the neighbourhood of the original values, will make it much more difficult, if not impossible, to identify the caller and also allows us to keep all the data, albeit with some noise, available for analysis.

In this section and the subsequent ones in this chapter, we will look at some of the ways to introduce calibrated noise to the data that will allow us to produce very accurate statistics while keeping the data private.

The best thing about differential privacy is its flexibility. Adding noise to the data won't make it easily identifiable, reduce the number of rows or lead to major loss of information. In case of differential privacy, a small amount of noise is added to the required data in order to mask the information and prevent any individual identity to be revealed. This noisy input is then used for any future analysis or ML training.

Let's look at the maths involved in the differential privacy. A randomized algorithm f provides (ε, δ)-differential privacy if, for all neighbouring databases D_1 and D_2, and for any set of outputs S

$$P\big[f(D_1) \in S\big] \le e^\varepsilon \times P\big[f(D_2) \in S\big] + \delta$$

where, f is any function

e^ε is very close to 1, with $\varepsilon > 0$
large e^ε gurantees no privacy
ε (epsilon) is a privacy parameter
or the privacy budget (lower ε, stronger privacy)
δ is failure probability or tolerance.
If $\delta = 0$, then we have $(\varepsilon, 0)$ differential privacy

The notion of the sensitivity of a function is central to the design of differentially private algorithms. Given a function f on a dataset D_1, the sensitivity is used to adjust the amount of noise required for $f(D_1)$. If f is a function that maps a dataset (in a matrix form) into a fixed-size vector of real numbers, we can define the generic sensitivity of f as

$$\text{Sensitivity} = \delta(f) = \max_{D_1, D_0} \| f(D_1) - f(D_2) \|_i$$

Where $\|.\|_i$ denotes to l_i norm with $i \in \{1, 2\}$
D_1, D_2 are two neighbouring dataset

Sensitivity is the maximum change in output with a change in one record between D_1 and D_2. Alternatively, it is the maximum difference one row can have as we apply the differential privacy algorithm. The noise added to the D_1 to make it D_2 can be any noise of choice be it Gaussian or Laplace. In case of the function being a machine learning model, D_1 and D_2 are two datasets that differ by one training sample, and the sensitivity would be the maximum change in the output when one training sample is removed from the training set. The noise added in D_1 can be computed using the sensitivity and the privacy budget on either Laplace or Gaussian distribution.

Using Laplace:

$$\text{Private}\,\mathbf{R} = f\left(D_1\right) + lap\left(\frac{\delta f}{\varepsilon}\right)$$

Where δf denotes the sensitivity of f and Lap $(\frac{\delta f}{\varepsilon})$ represents the noise drawn from the Laplace distribution with the centre of 0 and the scaling of $\frac{\delta f}{\varepsilon}$

Using Gaussian:

$$\text{Private}\,\mathbf{R} = f\left(D_1\right) + N\left(0,\,\sigma^2\right)$$

Where $N(0,\,\sigma^2)$ indicates that the noise variable is independently and identically distributed Gaussian distribution with the standard deviation $\sigma = \delta_2\left(f\right) \times \dfrac{\sqrt{2\log\dfrac{1.25}{\delta}}}{\varepsilon}$, where $\delta_2(f) = \|f(D_1) - f(D_2)\|_i$. Unlike k-anonymity, DP is data-agnostic and depends on privacy budget and the distribution being used. As long as the differentially private algorithm is being used, the resulting data will be differentially private too.

As discussed above, ε (epsilon) is the privacy budget which decided the amount of privacy in the algorithm. A small value would ensure high privacy, while high value of ε provides less privacy (you can afford to give away more of the privacy when you have more budget). It is advisable to have ε (epsilon) less than or equal to 1.

Let's consider couple of examples for different query functions and the impact of different values of ε on the value generated after the DP treatment. In the code example below, we are adding Laplace noise to a query fetching number of instances where the Applied Amount is greater than 2300. The subsequent plot shows the generated values across the 100 loops. The dotted black line is the actual value, and the different colours correspond to the value of ϵ. From the chart (Fig. 8.1), we can see that the privacy varies as we vary the privacy parameter. However, for a given epsilon, the value returned is very close to actual value, thus preserving differential privacy.

```
sensitivity = 1
#epsilon = 0.1

original = X_train[X_train['AppliedAmount'] >= 2300].shape[0]

V=[]
ep=[]
count=[]

for j in range (1,6):
    epsilon = j/10

    for i in range(0, 101):

        value = X_train[X_train['AppliedAmount'] >= 2300].shape[0] + \
        np.random.laplace(loc=0, scale=sensitivity/epsilon)
        V.append(value)
        ep.append(epsilon)
        count.append(i)
```

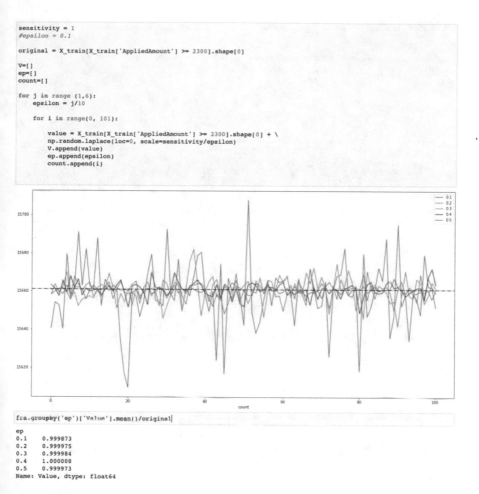

```
fra.groupby('ep')['Value'].mean()/original
```

```
ep
0.1    0.999873
0.2    0.999975
0.3    0.999984
0.4    1.000008
0.5    0.999973
Name: Value, dtype: float64
```

Fig. 8.1 The chart shows how the ε affects the values generated

In another example (Fig. 8.2), we have a function to compute the mean of Applied Amount (for Applied Amount greater than 2300). Here the sensitivity is calculated as a maximum difference if one record is added with maximum possible Applied Amount. The code adds return the mean value which is differentially private.

```
sensitivity = ((X_train['AppliedAmount'][X_train['AppliedAmount'] >= 2300]).max())/(len(X_train)+1)
print(sensitivity)
#epsilon = 0.1

original = (X_train['AppliedAmount'][X_train['AppliedAmount'] >= 2300]).mean()

V=[]
ep=[]
count=[]

for j in range (1,6):
    epsilon = j/10

    for i in range(0, 101):

        noised = (X_train['AppliedAmount'][X_train['AppliedAmount'] >= 2300]).mean() + \
        np.random.laplace(loc=0, scale=sensitivity/epsilon)
        value=noised
        V.append(value)
        ep.append(epsilon)
        count.append(i)
```

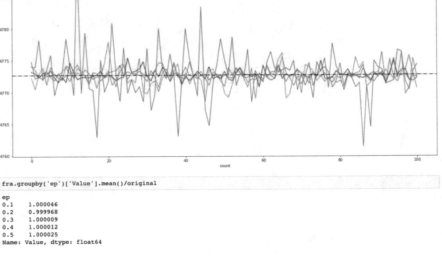

```
fra.groupby('ep')['Value'].mean()/original
```

```
ep
0.1    1.000046
0.2    0.999968
0.3    1.000009
0.4    1.000012
0.5    1.000025
Name: Value, dtype: float64
```

Fig. 8.2 The output reveals that in most of the cases the noise outcome is very close to the original mean with epsilon value of 0.3 returning the most closest value

Privacy Using Exponential Mechanism

The previous section shows how the data can be made differentially private by adding noise. However, adding noise to the data that is non-numeric in nature can give some really weird results. For instance, adding noise to a zip code can result in a value that may not exist or adding noise to an integer that represent a category may not work too. Also, it won't work for binary or Boolean data type and is not suitable for features that are categorical in nature like colour of a car, gender, marital status and so on. For these cases, a slightly different mechanism needs to be adopted to introduce noise.

Exponential mechanism (EM) aims at adding noise to a function that is dealing with non-numeric features. The technique we will cover in this section will add noise, to satisfy privacy, by sometimes returning a false value. It will, however, always return a value that is part of the dataset and not a value outside of the data universe on which the EM is implemented. This helps improve privacy and also eliminates the chances of having invalid values in the dataset.

It would select the element from a set of elements given a scoring function. The scoring function can be used to find K-top value (or K-bottom) for the field from the data, like most common zip code, the least preferred colour of a car, gender of people who are employed for more than 10 years in a company or marital status of patients in a given hospital. In order to satisfy privacy, exponential mechanism would sometime return a false number. For instance, it can return a colour of a car that is most preferred rather than least preferred.

Consider a categorical data (x) with values in the range (R) with a scoring function (u). The sensitivity of the scoring function thus would be

$$\delta u = \max \| u(x,r) - u(y,r) \|$$

EM $(M_E(x, u, R))$ would return a value $r \in R$ with probability proportional to $\exp\dfrac{(\varepsilon u(x,r))}{\delta u}$. It can be also used in conjunction with differential privacy by adding Laplace noise $(\dfrac{\delta u}{\epsilon})$ to the score function (u) and returning the value that has maximum noise but falls in the dataset.

For instance, in the dataset, the rating variable has eight categories with rating C and D having maximum frequency while F and HR having least frequency.

```
datafile['Rating'].value_counts()/datafile.shape[0]

C     0.244160
D     0.205170
B     0.196178
E     0.169985
A     0.084515
AA    0.041652
F     0.033478
HR    0.024861
```

The below code aims at returning one category out of the total of eight for the rating feature that has the highest frequency. The code declares the score function followed by a function that is used to return differentially private outcome by adding Laplace noise to the outcome of the Score function. Here the Score function is about returning the proportion of the categories in a given feature; here, for instance, it would return the proportion of the different ratings from the rating variable of the dataset followed by Laplace max function to return the rating categories with highest proportion in data albeit with noise.

```
def score(data, option):
    return data.value_counts()[option]/data.shape[0]


def laplace_max(df, values, scoref, sensitivity, epsilon):
    # Calculate the score for each element of R
    scoresf = [scoref(df, val) for val in values]

    lap_noise = [(scoref + np.random.laplace(loc=0, scale=sensitivity/epsilon)) for scoref in scoresf]

    idx = np.argmax(lap_noise)

    return values[idx]
```

```
values = datafile['Rating'].unique()
```

```
laplace_max(datafile['Rating'], values, score, 1, 1)
```

```
'A'
D       0.17
A       0.15
C       0.14
AA      0.14
F       0.12
HR      0.11
E       0.09
B       0.08
..       --
```

The above code returns A if executed once but would return D if iterated for infinite times.

Similarly, the below set of code used EM method probability function and returns C as the value with highest frequency.

```
def exponential(df, values, scoref, sensitivity, epsilon):
    # Calculate the score for each element of R
    scoresf = [scoref(df, val) for val in values]

    # Calculate the probability for each element, based on its score
    probabilities = [np.exp(epsilon * scoref / (2 * sensitivity)) for scoref in scoresf]

    # Normalize the probabilties so they sum to 1
    probabilities = probabilities / np.linalg.norm(probabilities, ord=1)
    #print(probabilities)

    # Choose an element from R based on the probabilities
    return np.random.choice(values, 1, p=probabilities)[0]
```

```
exponential(datafile['Rating'], values, score, 1, 1)
```

```
'C'
```

The choice of the method largely depends on the size of the range. In case the returned value needs to be from finite set of data (credit rating), EM would be advisable, but in case the range is infinite, or the labels are numerically coded (e.g. name of a person or post code), EM with Laplace can be opted.

Differentially Private ML Algorithms

So far, we have looked at the ways to add differential privacy to the data. Adding calibrated noise to the data allows us to then use any algorithm we want to train the model and still achieve the data privacy. However, the concept of differential privacy can be extended beyond adding the noise to the data.

In this section, we will see how we can use similar approach to add differential privacy to the model, by adding noise to the model's objective function, or to the model output, by adding noise to the weights of the output of the model's objective function.

Privacy can be introduced at input data (using noise or any other techniques), add noise to the objective function that satisfy privacy or add noise derived from Laplace or Gaussian distribution to the weights of the output of the objective function. Adding noise to the objective or the weights of the output has a similar effect as adding noise to the training data, with an added protection against any attack on the model.

Here's a standard form of classification objective function.

$$J(W) = argmin \; \frac{1}{n} \sum_{i=1}^{n} l\left(W, (x_i, y_i)\right)$$

For input perturbation (adding noise to the input data), the objective functions become

$$J(W) = argmin \; \frac{1}{n} \sum_{i=1}^{n} l\left(W, (x_i + z, y_i)\right)$$

In case of output perturbation (adding noise to the output weights), the objective functions become

$$J(W) + (W, Z)$$

where Z is noise derived from Laplace or Gaussian distribution and has the same dimensions as W. The noise gets added to the weights after the training is complete to make the output differentially private.

However, when we want to add the privacy to the objective function, we skew the function by adding noise (derived from Laplace distribution) to coefficient of the model. In case of tree-based algorithm, noise would be added to information gain parameter.

Let's consider dataset $D = \{(x_i, y_i)\}$ and objective function.

$$J(W) = argmin \; \frac{1}{n} \sum_{i=1}^{n} l\left(W, (x_i, y_i)\right)$$

The loss function after privacy parameter has been added in the objective function will be

$$J(W) = argmin \; \frac{1}{n} \sum_{i=1}^{n} l\big(W, (x_i, y_i)\big) + (W \times Z)$$

Where Z is the noise vector

As discussed above, the noise added depends on the sensitivity and privacy parameter. Since we are updating the objective function, implementing these techniques would mean writing the algorithm from scratch or use a privacy-preserving version of an algorithm.

Another way to do this is by adding noise to the gradient by optimizing SGD with noisy gradient. SGD processes one data point at a time with incremental gradient steps. For this example, we will assume data $D = \{(x_i, y_i)\}$, with n records with $y_i \in \{0, 1\}$. Given an objective function, SGD can be used to find a separating hyperplane. Here SGD can be defined as

$$W_{t+1} = W_t - \eta\big(\delta l\big(y_t, W_t, x_t\big) + \lambda W_t\big)$$

where η is the learning rate and gradient is computed on single instance (y_t, x_t)
In case of SGD mini batch, the weights would be computed as

$$W_{t+1} = W_t - \eta\left(\frac{1}{b}\sum \delta l\big(y_t, W_t, x_t\big) + \lambda W_t\right)$$

where b is the size of the mini – batch

By adding a noise parameter, SGD can be made differentially private, and thus the weights would be calculated as

$$W_{t+1} = W_t - \eta\big(\delta l\big(y_t, W_t, x_t\big) + \lambda W_t + Z_t\big)$$

Where Z is the noise vector

And in case of mini-batch SGD, the equation would slightly change to

$$W_{t+1} = W_t - \eta\left(\frac{1}{b}\sum \delta l\big(y_t, W_t, x_t\big) + \lambda W_t + \frac{1}{b}Z_t\right)$$

As noise is being added at each iteration of gradient decent, each data point would have different value (parameter of differential privacy) of differential privacy. In case of SGD or mini-batch SGD, privacy implementation would be dependent on batch size and learning rate.

Note: How Do Leaders in Tech Implement Privacy?
Apple uses local differential privacy where it is computed on individual devices before being collected by the central server. It also sets a limit on the frequency of data that can be collected from a particular user. Data is privatized using DP on user device for most of the events, like typing an emoji, and is stored on the device itself. Later, a random sample of this differentially private data is later transmitted to the server. Another DP technique called private count mean sketch (CMS) is used, where instead of sending the data as-is, a histogram of differentially private counts is sent to the server, using variations of SHA-256 hash where the input is encoded as a vector, followed by each coordinate of the vector being flipped with a probability of $\dfrac{1}{1+e^{\varepsilon}}$ to ensure privacy of the transmission, where ε is the privacy parameter.

Google has a lot of associations with third party for sharing navigation data. In order to maintain privacy, it shares anonymized aggregated randomly sampled historical traffic statistics (like average traffic speed, relative volumes and trajectory). The anonymized aggregated is differentially privatized by noise before data transmission.

Microsoft has also developed its own local DP mechanisms for collection of counter data for their basic analytical tasks, i.e. mean and histogram estimation. The mechanism is used by Microsoft to collect telemetry across devices. In this case, when the collection of counters $x_i(t)$ at time t_i is requested, the user (i) transmits one bit $b_i(t)$ which is sampled from a distribution where the user is either 0 *or* m, using which the mean can be computed.

At LinkedIn, differential privacy is applied on the user data before an analyst gets access to it. Using local differential private algorithm, a deterministic, pseudorandom noise – from the appropriate Laplace distribution – which is custom to a query is implemented.

Federated Learning

In the examples we have seen so far, the models are being trained locally with all the training data available to them. This makes the data available to the teams involved even before it is made differentially private and can be a challenge in sharing the sensitive data. In many cases, sharing the sensitive data can be a challenge even after it has been made differentially private as the owners of the data may not accept their data being sent to a central data warehouse. This is where federated learning comes in. It allows the owner of the data to allow the system to utilize the data for insights without sharing the actual data.

In federated learning, the data remains on the remote system where the ML algorithms train and then the relevant output is transmitted to the central system. This approach is extremely useful in domains where user data privacy is very important, and a data breach can have legal/regulatory implications (however, it is also a good practice and is gaining more ground). When used in healthcare, for example, the ML model can be trained on the user device or the remote location, and the updated model parameters from the remote machine can be used by the central model to update itself without any privacy concerns. The final model, which contains insights extracted from a large and varied dataset, can then be sent back to the individual's machine.

In order to further secure the system, data or model being trained locally can be made differentially private (by adding noise to input, objective function, gradient or output) before being shared with the central model.

Take, for instance, a system with K nodes, P_k ($k \in 1...K$) being the dataset on each node. The size of the data on node k is $n_k = |P_k|$ with n being the total private data across all nodes. Thus, model weights would be computed as

$$f(w) = \frac{n_k}{n} \sum_{1}^{K} F_k(w)$$

And the local loss function would be

$$F_k(w) = \frac{1}{n_k} \sum_{P_k} f_i(w)$$

With highly distributed data collection from the mobile devices, federated learning has been adopted by the tech majors as a way to continue to improve their products while making their products better. This also allows them to provide more customized experience to the users at times. For example, Google uses federated learning approach for improving the keyboard experience where the word suggestions model resides on the user's device and is trained on user typing pattern. The model parameters are sent to Google's central server which is then used for developing a master model (along with parameters from models trained on data of millions of other users) and then sent back to the devices for better predictive text feature. Model parameters are also made differentially private before they leave user device.

Apple also uses combination of differential privacy and federated learning for many of its features that is available on Apple's device. Combination of DP and FL is used for predictive keyboard and error correction and for Siri.

Note: Privacy Preserving Fairness

In previous chapters, we have seen how to detect and mitigate bias across various stages of data science lifecycle. We have also tried connecting fairness, explainability and monitoring with each other and have seen their role in ethical AI at large. Let's now talk about how we can combine privacy and fairness together to create a privacy preserving fairness approach. Ensuring fairness includes removing the biases that are present in the data, finding the relationships that independent features have with sensitive features and reducing the impact of those relationships. Since privacy is adding noise in the data, we would expect it to weaken those relationships automatically and hopefully make the algorithm fairer by default. Let's look at how we achieve that.

In a method called PDFC (purely and approximately differentially private and fair classification), differential privacy is used to ensure fairness. A protected feature (S) is randomly selected, and noise is added to the co-efficient (weights), thus reducing its correlation with other unprotected features (bringing in fairness) but also protecting the feature from any attack by adding noise (and hence adding privacy). The PDFC algorithm is quite intuitive. The privacy budget ε is decided separately for protected (S) and non-protected (X) features as ε_S and ε_X. The noise is added to the parameters/weights as per their assignments, and the objective function is "reconstructed" using the DP-objective function. However, the twist here is that the objective function does not only have noise added but also a fairness constraint to satisfy the fairness metric. The fairness constraints are added as a regularizer to the linear classifier loss function.

Choice of Regularizer

One of the regularizer that can be added in the loss function of a linear classifier is the decision boundary fairness constraint.

$$g_1(x,\theta) = \left| \sum_{i=1}^{n} (S_i - \overline{S}) \times \theta \right|$$

where $\overline{S}$ is average value of the protected data
θ is the optimization parameter

Here the regularizer will penalize the loss function if the distance between a single observation of a protected feature and average value of the protected feature is high. In case of no discrimination where $S_i - \overline{S} = 0$, the penalization would be zero.

Another regularizer that can be used is

$$g_2(x,\theta) = \sum_{i=1}^{n} \left(\theta_s \left[\log \tau - \log \phi \right] + (1 - \theta_s) \times \left[\log(1-\tau) - \log(1-\phi) \right] \right)$$

where $\theta_s = P(Y = 0 \mid S, x)$; $1 - \theta_s = P(Y = 1 \mid S, x)$

$\tau = P(Y = 0 \mid S)$; $1 - \tau = P(Y = 1 \mid S)$
$\phi = P(Y = 0)$; $1 - \phi = P(Y = 1 \mid S)$

Here the regularizer will penalize the loss function if there is a huge difference in probability ($\log[P(Y = 0 \mid S)] - \log[P(Y = 0)]$) of outcome with ($\tau$) and without ($\phi$) the protected feature. In case the probability is the same ($\tau = \phi$) thus showing no discrimination, the penalization would be zero.

The noise added here can be from Laplace (or Gaussian) distribution with respective parameters as discussed above in the chapter. The sensitivity can be calculated as the max distance in the data if one training sample is removed.

The loss function would be defined using differentially private parameters separately for protected features and non-protected features (different noise added to the parameters for protected and non-protected features) and then would minimize the whole loss function with fairness regularizer.

The point to note here is that a standard out-of-box machine learning algorithm is not likely to provide you with the flexibility to take these actions. To be able to use PDFC, you would need to create a custom ML model where the objective function is re-constructed to preserve privacy or fairness or even both of them.

Conclusion

Data privacy can often be an overlooked aspect when creating a machine learning algorithm. The basic data hygiene of replacing the original identifiers was considered as sufficient to bring in privacy until not too long ago. However, with the ubiquitous data collection around us, extracting private information from a dataset that does not have privacy built into it is now easier than ever. And as you build your product, you need to take strong measures to ensure privacy. It is important to note that as you work on adding privacy to your data, the parameters need to be selected carefully to avoid adding excessive noise to the data and have an adverse impact on the model parameters. The excess noise can cause one or more problems: exaggerated discrimination, reduced explainability or even faster model decay.

We have kept this as the last chapter in the book since the privacy impacts all the major themes covered in the book and covering them in detail before talking about privacy can help emphasize the impact. However, it is imperative to formulate the privacy strategy even before you create the ML approach. Adopting privacy in your ML project will not only offer an inherent protection for the user data, but it will also have a cascading impact of making it easier to reduce bias from your model and improve immunity against any attack on your model.

Bibliography

ADVANCED COMPOSITION (no date) Privacy preserving machine learning, Inria.fr. Available at: http://researchers.lille.inria.fr/abellet/teaching/ppml_lectures/lec3.pdf (Accessed: April 23, 2021).

Apple Machine Learning Research. 2021. Learning with Privacy at Scale. [online] Available at: <https://machinelearning.apple.com/research/learning-with-privacy-at-scale> [Accessed 16 April 2021].

Cs.utexas.edu. 2021. "How to Break Anonymity of the Netflix Prize Dataset" - FAQ. [online] Available at: <https://www.cs.utexas.edu/~shmat/netflix-faq.html> [Accessed 16 April 2021].

Cseweb.ucsd.edu. 2021. [online] Available at: <http://cseweb.ucsd.edu/~kamalika/pubs/WIFStutorial.pdf> [Accessed 16 April 2021].

Data Protection (no date) Europa.eu. Available at: https://edps.europa.eu/data-protection_en (Accessed: April 23, 2021).

Data Protection (no date) Europa.eu. Available at: https://edps.europa.eu/data-protection en (Accessed: April 23, 2021).

Ding, B., Kulkarni, J. and Yekhanin, S. (2017) "Collecting telemetry data privately," arXiv [cs. CR]. Available at: http://arxiv.org/abs/1712.01524.

Ding, J. et al. (2020) "Differentially private and fair classification via calibrated functional mechanism," arXiv [cs.CR]. Available at: http://arxiv.org/abs/2001.04958.

Dwork, C. (2008) "Differential privacy: A survey of results," in Lecture Notes in Computer Science. Berlin, Heidelberg: Springer Berlin Heidelberg, pp. 1–19.

Dwork, C. and Roth, A. (2013) "The algorithmic foundations of differential privacy," Foundations and trends in theoretical computer science, 9(3–4), pp. 211–407.

Dwork, C. et al. (2017) "Exposed! A survey of attacks on private data," Annual review of statistics and its application, 4(1), pp. 61–84.

Dwork, C., McSherry, F., et al. (2017) "Calibrating noise to sensitivity in private data analysis," Journal of Privacy and Confidentiality, 7(3), pp. 17 51.

Engineering.linkedin.com. 2021. Privacy-preserving analytics and reporting at LinkedIn. [online] Available at: <https://engineering.linkedin.com/blog/2019/04/privacy-preserving-analytics-and-reporting-at-linkedin> [Accessed 16 April 2021].

Gong, M. et al. (2020) "A survey on differentially private machine learning [review article]," IEEE computational intelligence magazine, 15(2), pp. 49–64.

Google Europe Blog. 2021. Tackling Urban Mobility with Technology. [online] Available at: <https://europe.googleblog.com/2015/11/tackling-urban-mobility-with-technology.html> [Accessed 16 April 2021].

Hern, A. (2021) "WhatsApp loses millions of users after terms update," The guardian, 24 January. Available at: http://www.theguardian.com/technology/2021/jan/24/whatsapp-loses-millions-of-users-after-terms-update (Accessed: April 23, 2021).

Ico.org.uk. 2021. Privacy attacks on AI models. [online] Available at: <https://ico.org.uk/about-the-ico/news-and-events/ai-blog-privacy-attacks-on-ai-models/> [Accessed 16 April 2021].

Ilvento, C. (2019) "Implementing the exponential mechanism with base-2 Differential Privacy," arXiv [cs.CR]. Available at: http://arxiv.org/abs/1912.04222.

Jagielski, M. et al. (2018) "Differentially Private Fair Learning," arXiv [cs.LG]. Available at: http://arxiv.org/abs/1812.02696.

Kamishima, T. et al. (2012) "Fairness-aware classifier with prejudice remover regularizer," in Machine Learning and Knowledge Discovery in Databases. Berlin, Heidelberg: Springer Berlin Heidelberg, pp. 35–50.

Narayanan, A. and Shmatikov, V. (2008) "Robust DE-anonymization of large sparse datasets," in 2008 IEEE Symposium on Security and Privacy (sp 2008). IEEE.

NPR (no date) "Cookie Consent and Choices." Available at: https://www.npr.org/2021/04/09/986005820/after-data-breach-exposes-530-million-facebook-says-it-will-not--notify-users (Accessed: April 23, 2021).

Pujol, D. et al. (2020) "Fair decision making using privacy-protected data," in Proceedings of the 2020 Conference on Fairness, Accountability, and Transparency. New York, NY, USA: ACM.

Rigaki, M. and Garcia, S. (2020) "A survey of privacy attacks in machine learning," arXiv [cs.CR]. Available at: http://arxiv.org/abs/2007.07646.

Song, S., Chaudhuri, K. and Sarwate, A. D. (2013) "Stochastic gradient descent with differentially private updates," in 2013 IEEE Global Conference on Signal and Information Processing. IEEE.

Sweeney, L. (no date) Simple demographics often identify people uniquely, Dataprivacylab.org. Available at: https://dataprivacylab.org/projects/identifiability/paper1.pdf (Accessed: April 23, 2021).

Tang, J. et al. (2017) "Privacy loss in Apple's implementation of differential privacy on MacOS 10.12," arXiv [cs.CR]. Available at: http://arxiv.org/abs/1709.02753.

TM Forum Inform. 2021. Google expands urban mobility project - TM Forum Inform. [online] Available at: <https://inform.tmforum.org/news/2015/11/google-expands-urban-mobility-project/> [Accessed 16 April 2021].

Truong, N. et al. (2020) "Privacy Preservation in Federated Learning: An insightful survey from the GDPR Perspective," arXiv [cs.CR]. Available at: http://arxiv.org/abs/2011.05411.

Truong, N. et al. (2020) "Privacy Preservation in Federated Learning: An insightful survey from the GDPR Perspective," arXiv [cs.CR]. Available at: http://arxiv.org/abs/2011.05411.

Uvm-plaid.github.io. 2021. Introduction — Programming Differential Privacy. [online] Available at: <https://uvm-plaid.github.io/programming-dp/intro.html> [Accessed 16 April 2021].

Veale, M., Binns, R. and Edwards, L. (2018) "Algorithms that remember: model inversion attacks and data protection law," Philosophical transactions. Series A, Mathematical, physical, and engineering sciences, 376(2133), p. 20180083.

What is considered personal data under the EU GDPR? (2019) Gdpr.eu. Available at: https://gdpr.eu/eu-gdpr-personal-data/?cn-reloaded=1 (Accessed: April 23, 2021).

Zafar, M. B. et al. (2015) "Fairness constraints: Mechanisms for fair classification," arXiv [stat.ML]. Available at: http://arxiv.org/abs/1507.05259.

Chapter 9
Conclusion

We hope that this book has given you a good understanding of issues you need to watch out for and the actions you need to take while developing a responsible AI model. The majority of effort today towards making fair and explainable models is being done by the regulated industries. However, any AI model, as is inherent to its nature, performs complex tasks that are not possible even with a large and complex implementation of a rules-based engine. Therefore, given the impact that AI has on the lives it touches, the responsibility to build RAI lies with everyone building intelligent products. Additionally, as the impact of the AI-driven products grows, the user awareness and the media coverage around the positive (or negative) impact they have on our lives has grown as well. If you are building an AI-driven product and have not looked at how to make it more responsible, it is time you do so now.

Responsible AI Lifecycle

In our experience, the teams utilizing RAI are mostly practicing it sporadically across the DS lifecycle. Most of the time the work on RAI starts either post model development or, worse, after modelling is done. This is partly because of lack of standard RAI lifecycle and partly because of lack of awareness on how to do it.

To make your approach to RAI achievable, repeatable and ultimately successful, you need to integrate it in your standard DS lifecycle. The diagram in Fig. 9.1 shows how the high-level steps in the complete DS-RAI lifecycle should look like. The different steps within the DS lifecycle may have different importance for you depending on the problem you are solving. Similarly, in the DS-RAI lifecycle, you may need to go much deeper in some areas than others. For example, if you already have a model live, then you may want to modify just the approach towards monitoring and defining the metrics that you want to monitor.

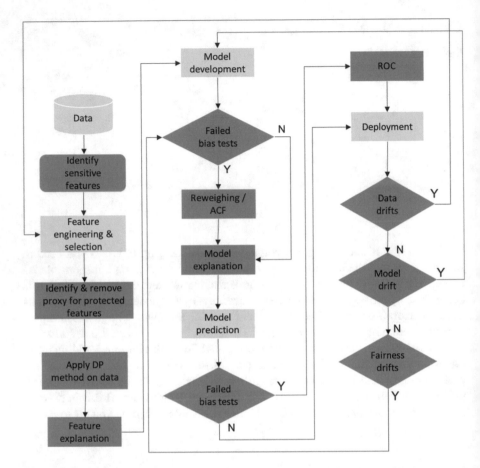

Fig. 9.1 A high-level flow of responsible AI across a data science lifecycle

Adding RAI to the DS lifecycle can make it more iterative as you may have to go back a few steps if you fail the bias tests. These iterations, as shown in the Fig. 9.1, will ensure that your final model is more robust and stands up to the challenges of responsible AI.

Responsible AI Canvas

Presented in Fig. 9.2 is the RAI canvas that can help ensure preparedness for creating a responsible model. In many cases, it may not be feasible to fill all the cards of the canvas on day one. In that case, any card left blank can be filled before you move past the stage in the lifecycle that requires it. As you are filling the canvas, capturing the acceptable thresholds for each metric will help understand if multiple cards are linked to each other.

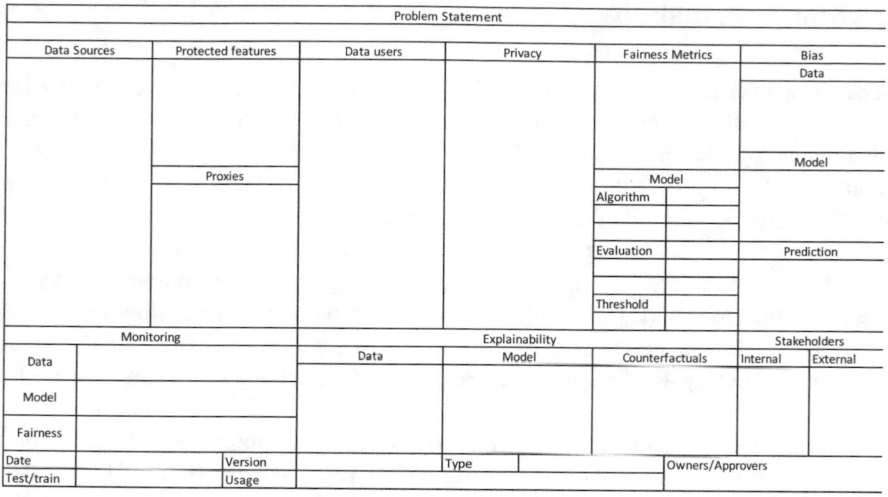

Problem Statement							
Data Sources	Protected features	Data users	Privacy	Fairness Metrics		Bias	
						Data	
	Proxies			Model		Model	
				Algorithm			
				Evaluation		Prediction	
				Threshold			
Monitoring		Explainability				Stakeholders	
Data		Data	Model	Counterfactuals		Internal	External
Model							
Fairness							
Date		Version	Type		Owners/Approvers		
Test/train		Usage					

Fig. 9.2 DS-RAI canvas with cards for all components of DS-RAI lifecycle

Note: Fair AI vs Responsible AI vs Ethical AI

Responsible AI term is often used as a replaceable term for fair AI and ethical AI, along with few other variants. Fair AI deals with fairness and is therefore a component of RAI. However, the ambit of RAI is just not confined to a fair AI but encompasses accountability, explainability and model privacy. It would be undeniable that XAI and AI monitoring are important to ensure right decisions for the customers and to ensure that none of our customers are penalized over time – which is an ethical principle.

The components of RAI (fairness, explainability, accountability and data and model privacy) in turn handle aspects that help make your model's behaviour more ethical. While ethical AI covers a wider set of parameters, the application of ethical AI when it comes to creating the model depends on RAI. Ethical AI deals with many more aspects of practicing AI and not just how the AI behaves with its users. The environmental impact, sustainability and the usage of AI are some other dimensions that ethical AI needs to address. Other components of ethical AI include usage for the deprived, human safety, UN sustainable development goals and more.

AI and Sustainability

The impact of AI is not confined to just the consumer-facing industries, it goes far beyond the realms of what a layperson knows and is impacting our lives in the ways that technology has never done before. Within the last decade, AI has been used to polarize the way people think to influence election results, create social credit score for the entire population of a country to control almost all aspects of their lives or use facial recognition to identify and prosecute legitimate protestors.

Singapore has recently started using AI to "engineer a harmonious society" – they also plan to use the technology to read citizen mood while developing policies for citizens. How far can such an experiment remain limited to positive implementation for the society and how long before this becomes another tool available to the Big Brother still remains to be seen.

The Financial Conduct Authority in the UK, which regulates around 60,000 financial services firms in the country, has collaborated with the Bank of England to create the Financial Services Artificial Intelligence Public-Private Forum (AIPPF) in order to understand the impact of algorithmic decision-making in FS sector. Clearly, the approach taken towards AI is very different across the world. While some countries are diving head in to implement AI with little regard to how its citizens would like to be governed, others are taking a more restrained approach and solving the ethical issues first.

In addition to the societal impact, the environmental impact of AI also deserves to be considered. The carbon footprint of the AI implementations has been a topic of many ongoing debates. For example, research shows that Google's AlphaGo Zero generated around 95 tonnes of carbon dioxide in 40 days during its model training. In comparison, an average British household generated about 8.1 tonnes of carbon dioxide in 2014. AlphaGo's training CO_2 emissions were more than that of eleven British households from 2014.

At the same time, AI is also being used to optimize energy consumption, better management of renewable energy sources, predicting energy consumption patterns and for optimizing energy needs for cooling data centres.

As more governments join the party and start using AI, the efforts should not be concentrated just to utilizing the technology to optimize the use of resources or reducing carbon footprint but should also focus on ethical use of these environmental factors. For example, AI can be used to reduce consumption and waste of energy but should further be used for equitable distribution of the energy. Similarly, monitoring marine pollution to improve health of the ecosystems and identifying oil spills using satellite imaging and AI will have an impact that'll go beyond reducing energy consumption.

The high carbon footprint of the AI algorithms posits another challenge. The cost of training the complex models is very high and is serving as an entry barrier for smaller businesses to adopt or build their own technology. Smaller businesses and in some cases even smaller countries cannot hope to compete with their larger and richer counterparts when it comes to building AI-driven society. While the internet and the related technologies had a great democratizing effect by giving the same power to build and launch tools to everyone, the high build cost of AI products is taking that away.

Need for an AI Regulator

As the RAI technology and research is moving on, the AI research is moving at an even faster rate. The new research into AI gives us more powerful tools to work with, but they also make it more challenging to implement RAI. While industry can remain on the cutting edge of the implementations, the grey area of what constitutes RAI or not will continue to widen as AI makes decisions related to almost everything in our lives – from whether an applicant is eligible for a mortgage, if a patient's scan shows cancer to which route you should take for your commute and which packet of peanut butter you buy based on the search results.

This problem is compounded even more for the regulated industries, where individual regulating bodies play catch up with the advancements in the technology. This would take us to a world where AI will be involved in making or influencing decisions related to almost everything in our lives – from whether an applicant is eligible for a mortgage, if a patient's scan shows cancer to which route you should take for your commute and which packet of peanut butter you buy based on the search results.

These challenges combined with those listed in previous sections create a strong case for an AI regulator. Traditionally regulators exist for an industry – like financial services or healthcare. However, AI is now fundamental to every industry, and individual regulators cannot be expected to keep up with the developments to update their guidelines or adoption of the technology. This can not only slow down the respective industries from adopting the latest developments but also deny consumers of the benefits.

A central AI regulator that focuses solely on the AI development and facilitating the adoption of the latest and greatest in the technology within the regulatory frameworks and parameters can serve as a subject matter expert for all other regulators and any unregulated industry that is using the AI to impact its users lives.

Recently EU came out with laws surrounding AI, to prevent irresponsible use that can harm society at large. In order to make AI safe, transparent, ethical and unbiased, the commission has made four risk categories:

- Unacceptable: "Anything considered a clear threat to EU citizens will be banned: from social scoring by governments to toys using voice assistance that encourages dangerous behaviour of children."
- High Risk: Use of AI in transport, education, employment, credit scoring, law, migration and justice falls into this category, wherein exhaustive checks need to be place before making AI live. There are few exceptions though:

- Conformity assessments, followed by registration in EU database and finally a declaration of conformity.
- Approval from notified bodies.
- Checks on quality of data to reduce discrimination.
- Traceability, documentation and transparency guidelines to be followed.
- Details to made available to consumers.
- High accuracy is must.
- Continuous monitoring.

- Limited Risk: Chatbots will fall into this category. Users need to be informed that the decisions (or their interaction) are machine based.
- Minimal Risk: Video games, spam filters and other use of AI will be tagged with minimal risk.

While this is a great step forward, given the massive impact of AI, more needs to be done to bring in the transparency in how the technology is being used. The decisions taken by the AI can have irreversible and life-altering impact on the users, and the transparency on how any business is using the AI and the public access to the RAI interface of their AI is key. One school of thought is to have companies issue an annual AI report. This is being explored by some regulators, but we believe providing transparency should be more in step with the development of technology and available much more on demand than once-a-year snapshot. Nevertheless, it will be a big step towards filling the void towards making AI more democratic and RAI more ubiquitous.

Bibliography

About the FCA (2016) Org.uk. Available at: https://www.fca.org.uk/about/the-fca (Accessed: April 27, 2021).

Excellence and trust in artificial intelligence (2020) Europa.eu. Available at: https://ec.europa.eu/info/strategy/priorities-2019-2024/europe-fit-digital-age/excellence-trust-artificial-intelligence_en (Accessed: April 27, 2021).

Harris, S. (2014) The Social Laboratory, Foreign Policy. Available at: https://foreignpolicy.com/2014/07/29/the-social-laboratory/ (Accessed: April 27, 2021).

Keramitsoglou, I., Cartalis, C. and Kiranoudis, C. T. (2006) "Automatic identification of oil spills on satellite images," Environmental modelling & software: with environment data news, 21(5), pp. 640–652.

Mohamadi, A., Heidarizadi, Z. and Nourollahi, H. (2015) "Assessing the desertification trend using neural network classification and object-oriented techniques (Case study: Changouleh watershed - Ilam Province of Iran)," İstanbul üniversitesi orman fakültesi dergisi, 66(2). doi: https://doi.org/10.17099/jffiu.75819.

Mozur, P. (2019) "In Hong Kong protests, faces become weapons," The New York times, 26 July. Available at: https://www.nytimes.com/2019/07/26/technology/hong-kong-protests-facial-recognition-surveillance.html (Accessed: April 29, 2021).

Org.uk. (No Date) Available at: https://www.theccc.org.uk/wp-content/uploads/2016/07/5CB-Infographic-FINAL-.pdf (Accessed: April 29, 2021).

Storm, D. (2015) ACLU: Orwellian Citizen Score, China's credit score system, is a warning for Americans, Computerworld.com. Available at: https://www.computerworld.com/article/2990203/aclu-orwellian-citizen-score-chinas-credit-score-system-is-a-warning-for-americans.html (Accessed: April 27, 2021).

van Wynsberghe, A. (2021) "Sustainable AI: AI for sustainability and the sustainability of AI," AI eand Ethics. doi: https://doi.org/10.1007/s43681-021-00043-6.

What artificial intelligence means for sustainability (no date) Greenbiz.com. Available at: https://www.greenbiz.com/article/what-artificial-intelligence-means-sustainability (Accessed: April 27, 2021).

Printed in the United States
by Baker & Taylor Publisher Services